T0349875

Probabilistic Lattices

With Applications to Psychology

ADVANCED SERIES ON MATHEMATICAL PSYCHOLOGY

Series Editors: H. Colonius (*University of Oldenburg, Germany*)
E. N. Dzhafarov (*Purdue University, USA*)

Published

Advanced Series on Mathematical Psychology Vol. 5

Probabilistic Lattices

With Applications to Psychology

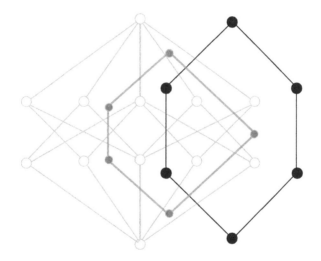

Louis Narens

University of California, Irvine, USA

World Scientific

NEW JERSEY · LONDON · SINGAPORE · BEIJING · SHANGHAI · HONG KONG · TAIPEI · CHENNAI

Published by

World Scientific Publishing Co. Pte. Ltd.

5 Toh Tuck Link, Singapore 596224

USA office: 27 Warren Street, Suite 401-402, Hackensack, NJ 07601

UK office: 57 Shelton Street, Covent Garden, London WC2H 9HE

Library of Congress Cataloging-in-Publication Data
Narens, Louis.
    Probabilistic lattices : with applications to psychology / by Louis Narens.
      pages cm -- (Advanced series on mathematical psychology ; vol. 5)
    Includes bibliographical references and index.
    ISBN 978-981-4630-41-2 (hardcover : alk. paper)
    1. Probabilities. 2. Probabilities--Psychological aspects. 3. Lattice theory. 4. Decision making--
Mathematical models. 5. Psychology--Mathematical models. I. Title.
    QA273.4.N372 2015
    519.2--dc23
                                                    2014038123

British Library Cataloguing-in-Publication Data
A catalogue record for this book is available from the British Library.

Printed in Singapore

For Kimberly and Beau

Acknowledgments

I want to thank Benjamin Feintzeig, Lisa Guo, and Sarita Rosenstock for many suggestions and corrections. They greatly helped to improve the quality of the text and the correctness of the mathematics.

The research for this book was supported by grants FA9550-08-1-0389 and FA9550-13-1-0012 from the Air Force Office of Scientific Research and grant SMA-1416907 from National Science Foundation.

Figure 6.2 was reprinted with permission from the *Journal of Mathematical Physics*. Copyright 1972, AIP Publishing LLC.

Contents

Chapter 1

Introduction

1.1 What This Book is About

This book is concerned with alternative forms of probability theory for philosophic and scientific application. Its emphasis is on understanding the limits of natural generalizations of standard probability theory, the relationship of these generalizations to logic, and their applications to science, especially to behavioral science. Very little attention is devoted to the classical issues in logic of deduction and completeness, and very little attention is devoted to methods of probabilistic inference and probabilistic laws.

In philosophy, alternative probability theories and logics abound. However, most of them are not applicable to science. In particular, their resulting numerical systems of probabilities generally lack sufficient algebraic structure to be useful in scientific applications. Two of these, however, appear to have much potential for scientific application.

One, "intuitionistic logic", started as a logic. It is the logic that is implicit in the mathematician L. J. Brouwer's theory of mathematics. Later researchers (e.g., Kolmogorov, 1933; Gödel, 1933) found other interpretations for it. Narens (2007) generalized it to provide a basis for a probabilistic theory of propositions that are either verifiable or refutable. He also used his generalization as a basis for developing concepts of "ambiguity", "vagueness", and "incompleteness" for the psychological modeling of human judgments of probability. This book provides a foundation of this generalization and applies it to rationality issues in the foundations of probability and decision making in behavioral economics.

The other alternative considered is known collectively in the literature as "quantum logic." It consists of algebraic generalizations of J. von Neumann's approach from the 1930s to the foundation of quantum physics. In this approach, von Neumann reduced logic to algebra. More precisely, he reduced logic to the study of a particular class of algebraic structures called "lattices" that had interpretations as closed subspaces of Hilbert space. A new theory of probability was developed for these lattices that shared many properties with traditional probability theory. Quantum logic treats these lattices both as "logics" and as domains for its version of "probability function." In recent years, it has had increasing scientific application outside of physics, including recent applications in the psychology of decision. For reasons of compactness, quantum logic as used in physics is not developed in this book. Instead, a close algebraic relative is formulated that shows promise for developing and applying new probabilistic concepts for decision theory and behavioral science.

Quantum and intuitionistic logics are natural extensions of classical logic and standard probability theory that retain many of the core properties of probability functions that have been traditionally used in scientific applications. Results presented throughout this book indicate that these two extensions, along with their natural variants, are likely the only interesting extensions that are applicable and useful for scientific application.

This first chapter describes the abstract approach to probability theory developed in later chapters. It is not meant as an overview of the material presented in those chapters, which contain much additional information. Instead, it is designed to show a progression of ideas and results from those chapters that leads to a generalized, applicable concept of "probability" and its relationship to logic. The intended and presented applications of the material of the book are to psychology and decision theory.

1.2   Level and Kind of Mathematics Used

This book is designed to be self-contained. Generally, technical concepts, including very common ones such as "boolean algebra of events", "finitely additive probability function", etc., are given explicit definitions, and usually theorems have complete proofs. There are exceptions: The explicit definitions of some concepts are not given when they are first encountered. Usually, these will be given later—sometimes in a later chapter. The same holds for proofs of some theorems. And a few theorems will be stated and used without proof. These will be either well-known results of mathematics or results from the literature with referenced proofs.

The main mathematical vehicle of the book is elementary lattice theory.

It is assumed that the reader may not be familiar with this part of algebra. Therefore, the needed parts of lattice theory are developed in a detailed manner. For the purposes of the book, parts of elementary lattice theory found in textbooks and the literature have to be modified, and some new results are needed.

The level of the mathematics employed is at the level of an upper division mathematics course at an American university. However, many of the ideas presented are sophisticated, and the principal kind of mathematics used (lattice theory) is likely to be unfamiliar to most readers, as well as some ideas from the foundations of mathematics (e.g., infinitesimals, the compactness theorem of logic). An extra effort is made to explicate these potentially unfamiliar ideas through the use of elementary proofs and examples.

1.3 How I Came to Write This Book

This book grew out of my research in decision theory. There, a number of experimental studies suggested that in making decisions, people employed features of uncertainty that were unaccounted for in standard probability theory. I decided to approach this issue analytically by generalizing standard probability theory to allow the measurement of the uncertainty of an event to take into account some of those features. This was accomplished by deleting a qualitative property of standard probability theory that I called "Binary Symmetry", defined as follows:

Binary Symmetry: Let A, B, C and D be mutually disjoint events. Let "$(A \,|\, A \cup B) \sim (B \,|\, A \cup B)$" stand for "the conditional probability of A given $A \cup B$ is the same as the conditional probability of B given $A \cup B$," and let $\sim$ be similarly defined for $(C \,|\, C \cup D) \sim (D \,|\, C \cup D)$, etc. Then the *Axiom of Binary Symmetry* says that if

$$(A \,|\, A \cup B) \sim (B \,|\, A \cup B) \text{ and } (C \,|\, C \cup D) \sim (D \,|\, C \cup D),$$

then

$$(A \,|\, A \cup B) \sim (C \,|\, C \cup D) \text{ and } (A \,|\, A \cup C) \sim (B \,|\, B \cup D). \quad \square$$

Empirical research shows that people take into account how well they understand the nature of the uncertainty in decision and probability estimation tasks. Consider a situation where for disjoint A and B a person judges $(A \,|\, A \cup B)$ as equally likely as $(B \,|\, A \cup B)$ based on a very good understanding of the natures of A and B, and for disjoint C and D judges $(C \mid C \cup D)$ just as likely as $(D \mid C \cup D)$ based on complete ignorance of the natures of C and D. Suppose the person is forced to choose between a \$100 bet of A occurring given $A \cup B$ and \$100 bet of C given $C \cup D$. The experimental literature suggests that many people strongly prefer A given $A \cup B$. The intuitive explanation for this is that most people are "ambiguity adverse," and prefer to gamble with known probabilities rather than ambiguous ones, all other things being equal. Some (e.g., Ellesberg, 1961) have considered this kind of ambiguity adversity to be consistent with rationality.

In Narens (2003) I axiomatized a version of qualitative conditional probability containing Binary Symmetry as an axiom. I deleted Binary Symmetry from the axiom system to obtain a generalization of conditional probability. The axioms for both systems were formulated in terms of the preference relation $\precsim$, where for $A \subseteq C$ and $B \subseteq D$, the relation "$(A \mid C) \precsim (B \mid D)$" is interpreted as "the event A given the event C is at least as likely as the event B given the event D." For the axiom system containing Binary Symmetry, it was shown that there existed a conditional probability function $\mathbb{P}$ such that for all conditional events $(A \mid C)$ and $(B \mid D)$ under consideration,

$$(A \mid C) \precsim (B \mid D) \text{ iff } \mathbb{P}(A \mid C) \le \mathbb{P}(B \mid D). \tag{1.1}$$

For the generalized system without Binary Symmetry, it was shown that Equation 1.1 generalized to the following: There exist a condition probability function $\mathbb{P}$ and a function v from events into the positive reals such that for all conditional events $(A \mid C)$ and $(B \mid D)$ under consideration,

$$(A \mid C) \precsim (B \mid D) \text{ iff } \mathbb{P}(A \mid C)v(A) \le \mathbb{P}(B \mid D)v(B). \tag{1.2}$$

In general, the interpretations of $\mathbb{P}$ and v in Equation 1.2 will vary with the intended application. The range of v will also depend on the intended application.

Call C in $(A \mid C)$ in Equation 1.2 the *conditioning event (of $(A \mid C)$)*, A the *focal event (of $(A \mid C)$)*, and $\mathbb{P}(A \mid C)v(A)$ the *belief of A given C*. $\mathbb{P}(A \mid C)v(A)$ is often written as $\mathbb{B}(A \mid C)$, and when so done, $\mathbb{B}$ is called a *belief function*.

Some theories of probability are based on the idea that uncertain events have a propensity to occur. This propensity, which is a qualitative idea, is taken to be 1-dimensional, and is measured quantitatively by a probability function. Some theorists believe that additional dimensions of uncertainty should also be considered in the measurement of belief, for example, dimensions like vagueness and ambiguity, and that these dimensions, along with probability, are relevant for some kinds of decisions involving uncertainty. Equation 1.2 provides a theory for measuring uncertainty in terms of probability and another dimension. In that equation, $\mathbb{P}$ measures the impact of conditional probability on the belief for the conditional event $(A \mid C)$ with $A \subseteq C$, while v measures the impact of the other dimension. v's impact is entirely determined by the focal event A and thus is independent of the conditioning event C.

I tried to use the belief function $\mathbb{B}$ to provide useful models of non-expected utility theory for economic and psychological decision making. I failed. For purely probabilistic situations, I had greater success. I found situations where it was natural to evaluate events in two ways, for example, the number of times an event theoretically occurred and the number of times it was observed to occur. I had the conditional probability function $\mathbb{P}$ computed in terms of one of these, for example, the number of times the event theoretically occurred, and had the function v be computed in terms of both kinds of occurrences, for example, the ratio of observed occurrences to theoretical ones. Additional research (Narens, 2003) revealed that the subset of events with empirically determined probabilities had a different logical structure from totality of events. For example, observable events do not, in general, form a boolean algebra of events, and because of this, the logical structure of observable events is, in general, different from the logical structure of events occurring in standard probability theory. This led to the consideration of extending belief theory *and* probability theory to collections of events that are in some cases richer or more general in structure than standard probability theory.

For the purpose of illustration, consider the case where "$\dagger\alpha$" is read as "α is verifiable". Then

$$\mathbb{P}(\alpha) \text{ is the probability that } \alpha \text{ is true,}$$

and

$$\mathbb{P}(\dagger\alpha) \text{ is the probability that it is true that } \alpha \text{ is verifiable.}$$

$\mathbb{P}(\dagger\alpha)$ and $\mathbb{P}(\alpha)$ are related as follows:

$$\text{always } \mathbb{P}(\dagger\alpha) \leq \mathbb{P}(\alpha) \text{ and often } \mathbb{P}(\dagger\alpha) < \mathbb{P}(\alpha).$$

Read "$\vdash\alpha$" as "α is refutable". Then

$$\mathbb{P}(\vdash\alpha) \text{ is the probability that it is true that } \alpha \text{ is refutable.}$$

Refutability can happen in a number of ways. Gödel (1931) provided an interesting example of one of these in his famous theorem about the incompleteness of arithmetic.

Example 1.1. Consider a formal axiomatization of arithmetic where for each proposition α, "$\neg\,\alpha$" stands for "not α", "$\dagger\alpha$" stands for "α is a theorem of the formal axiomatization", and "$\vdash\alpha$" stands for, "The assumption that α is a theorem leads to a contradiction," where the contradiction can occur by either $\neg\,\alpha$ being a theorem or by metamathematical consideration demonstrating that α is not a theorem, like, for example, the metamathematical considerations used in Gödel (1931) for establishing the incompleteness of arithmetic. It then follows from Gödel (1931) that for each considered sentence of arithmetic, α,

$$\mathbb{P}(\dagger\alpha) \leq \mathbb{P}(\vdash\vdash\,\alpha) \leq \mathbb{P}(\alpha)\,,$$

and there exists an arithmetic sentence β such that

$$\mathbb{P}(\dagger\beta) < \mathbb{P}(\vdash\vdash\,\beta) \leq \mathbb{P}(\beta)\,.$$

Note that for all arithmetic propositions α and β such that α implies β,

$$\mathbb{P}(\dagger\alpha\,|\,\beta) \;=\; \frac{\mathbb{P}(\dagger\alpha)}{\mathbb{P}(\beta)} \;=\; \frac{\mathbb{P}(\alpha)}{\mathbb{P}(\beta)}\cdot\frac{\mathbb{P}(\dagger\alpha)}{\mathbb{P}(\alpha)} \;=\; \mathbb{P}(\alpha\,|\,\beta)\cdot v(\alpha)\,, \tag{1.3}$$

where

$$v(\alpha) = \frac{\mathbb{P}(\dagger\alpha)}{\mathbb{P}(\alpha)}\,. \tag{1.4}$$

Also note v in Equation 1.4 is a function, because $\mathbb{P}(\dagger\alpha)$ and $\mathbb{P}(\alpha)$ are completely determined by α. The right-hand side of Equation 1.3,

$$\mathbb{P}(\alpha\,|\,\beta)\cdot v(\alpha)\,,$$

has the algebraic form of the previously discussed belief function $\mathbb{B}$ arising from a qualitative axiomatization for generalized conditional probability. Thus in this case, $\mathbb{B}(\alpha\,|\,\beta) = \mathbb{P}(\alpha\,|\,\beta)\cdot v(\alpha)$, where $v(\alpha)$ is the probability $\mathbb{P}(\dagger\alpha\,|\,\alpha)$. □

The operation $\vdash$ in Example 1.1 does not have the properties of the negation operation, *not,* of logic. The key property that it is missing is the Law of the Excluded Middle that requires for all propositions α, "either α or $\vdash\alpha$" is true (= certain). In Example 1.1, the failure of the Law of the Excluded Middle for β leads to $v(\beta) < 1$.

Intuitionistic Logic is a well-known generalization of classical logic in which the Law of the Excluded Middle fails. I decided to investigate decision theories based with probabilities defined on algebras of events corresponding to intuitionistic logic instead of on a boolean algebra of events corresponding to classical logic. I found that intuitionistic based event spaces had natural interpretations in experimental psychological experiments involving human judgments of probability while providing interpretations for the formula,

$$\mathbb{B}(\alpha \mid \beta) = \mathbb{P}(\alpha \mid \beta) \cdot v(\alpha),$$

for the degree of belief of α given β for α and β in the intuitionistic event space. This led me to look for other generalizations of boolean algebras of events that may be of use for modeling human decision making.

About the same time I was exploring the use of intuitionistic logic for probabilistic application in science, the psychologist Busemeyer, working with mathematicians and psychologists, used von Neumann's logic and probability theory for quantum mechanics for explaining experimental results in human decision making. Von Neumann's logic and its associated probability theory had different algebraic properties from the kind of logic and probability theory that I was developing. This led me to investigate the range of logics and probability functions that looked useful in experimental scientific applications, particularly applications in psychology. This book is the result of those investigations. It delves into the consequences of basing probability and belief on non-boolean algebras of events. Later parts of this chapter explore how the algebraic properties of functions used to measure uncertainty impact the logical structure of the underlying event space.

Quantum logic is the theory about empirical observations in quantum mechanics. The possible empirical observations are restricted by the logical structure of the experimental situation under consideration *and* the laws of quantum mechanics. Busemeyer and colleagues realized that many puzzling experimental results in decision theory arose from experiments that had the same logical structures of those of quantum mechanics. They explored the psychological ramifications of this by using the methods of quantum logic, and they were able to provide theoretical explanations of the puzzling results of the decision experiments. I thought that instead of using von Neumann's quantum logic, a related formulation more closely aligned with standard probability theory and experimental-psychological methodology could be used to capture the quantum-like nature of the decision phenomena that Busemeyer and colleagues were modeling. This program is carried out in Chapter 6, where it is shown that the logical structure of multi-condition psychological experiments can lead to non-boolean event structures, including ones isomorphic to those found in quantum mechanical experiments. Furthermore, a form of probabilistic non-additivity often results, in which the probability of the empirical disjunction of two disjoint events can be greater than the sum of their probabilities. This form of non-additivity is characteristic of physical quantum phenomena. Thus probabilistic non-additivity can be a direct result of the logical structure of the experimental paradigm. In quantum mechanics, additional forms of probabilistic non-additivity result from quantum mechanical laws. Similarly, for some forms psychological experimentation additional forms of probabilistic non-additivity can result from explicitly stated psychological laws. This is especially the case for experiments where a subject's participation in one condition of the experiment would influence her results in a different condition. The nonadditivity results from the use of laws linking the experimental conditions so that counterfactual reasoning of the following sort can be used: "If the subject did x in Condition A then she would have done y in Condition B."

1.4 Differentiation from Fuzziness

Some of the applications of the probability theories and logics developed in this book employ concepts of "ambiguous," "vague," and "poor" elements. Because applications of fuzzy set theory often use similar terminology to describe their phenomena, there exists potential to confuse this book's developments with some of those of fuzzy set theory. However, there is nothing "fuzzy" or even probabilistic about the concepts of "ambiguous," "vague," and "poor" used in the developments of this book, because they are defined in terms of clearly specified sets with a deterministic membership relation. In contrast, fuzzy set theory, as formulated in Zadeh (1965), has a probabilistic membership relation, that is, for each element b of the universe under consideration and each set B, there is a probability p that b is in B and a probability $1 - p$ that b is not in B. The fuzzy elements of B are those that have probability different from 0 or 1 of being in B.

Associated with a set B in fuzzy set theory are the elements of the universe under consideration that:

- clearly belong to B—the elements b such that b is in B with probability 1;

- fuzzily belong to B—the elements b such that b is in B with probability p, $0 < p < 1$; and
- clearly do not belong to B—the elements b such that b is in B with probability 0.

Metaphorically, the fuzzy elements form a boundary separating those that belong to B from those that belong to the set-theoretic complement (with respect to the universe under consideration) of B. This "boundary" is not a boundary in the topological sense of the word.

In contrast, as is shown in Chapter 3, many of the event spaces considered in this book can be viewed as open sets from a topological space with universe X, where each event has a topological boundary. (Definitions of basic topological concepts are provided in Chapter 3, Definitions 3.5 and 3.6.) "Ambiguous", "vague", "poor", and other concepts relevant to human judgments of uncertainty are then defined in Chapter 5 through the use of properties of topological boundaries.

1.5   Toward a Generalized Theory of Probability

Probability theory is designed to measure the uncertainty in events. In its standard form, the events form a boolean algebra and their measurement is accomplished through a normed, countably additive measure function. Such a measurement system allows the powerful methods of analysis from mathematics to be applied, and much of the use of probability in physical science, engineering, and statistics rely on the powerful tools provided by this system. For behaviorally based decision science, probability theory is often generalized from a countably additive measure to finitely additive measure, resulting in a loss of much deep mathematics. However, finite additivity is easier to justify philosophically *for the measurement of uncertainty,* and it allows for the incorporation of some interesting decision concepts that are outside of standard probability theory. Because of this book's emphasis on psychologically motivated decisions, its probability theory is formulated in terms of finite additivity, although much of its mathematical theory could be extended to include the more standard, countable form of additivity.

The usual approach for generalizing probability theory in the literature is to generalize the algebraic form of the probability function. However, a principal use of probability theory in science and applications is its calculative power, and such generalizations usually greatly weaken calculative power, making for a more general but a mathematically poorer and scientifically less applicable theory of probability. The approach of this book is very different: It keeps the algebraic form of the probability function, but changes the algebra of events. The particular changes in event algebra employed allow the generalized probability function to retain most of its calculative power, while allowing for a more general, richer, and deeper concept of uncertainty.

Historically, a change in event structure for probability theory occurred in the 1930s when von Neumann developed his theory of probability for quantum mechanics. Von Neumann's probability theory exploited the richer mathematical structure of the event space consisting of closed subspaces of a Hilbert space. This allowed him to develop a probability theory that was appropriate for phenomenon of entanglement observed in quantum physics. Chapter 6 presents an alternative approach to probability theory that has many characteristics in common with von Neumann's approach. The alternative, which can be interpreted as a fragment of standard probability theory, is based on a very different set foundational ideas than those of von Neumann.

1.6   Extended Probability Functions

Modern probability theory is based on the mathematics of measure theory, which historically was about the concept of geometric volume (which includes length and area). For such a geometric development, it is reasonable to consider sets of measure zero as "having no volume". For probability theory, I consider it odd to think of a non-null event of probability zero as having "no probability", because such an event can still occur. This oddity results from the Archimedean property of the real number system, which is used to measure "probability". This oddity can be made to disappear through the use of extensions of the real numbers that include positive infinitesimals.

Beyond the philosophical issue involving the occurrences of non-null events of measure zero, there is a more important reason for extending probability functions to include positive infinitesimal values: The resulting mathematical theory is richer, more applicable to decision theory, and the proofs of many important theorems of standard probability theory become much easier.

Convention 1.1. *Throughout the book, the following notation, conventions* *and definitions involving the real numbers and sets are observed:* $\mathbb{R}$ denotes the set of reals, $\mathbb{R}^+$ the set of positive reals, $\mathbb{I}$ the integers, $\mathbb{I}^+$ the positive integers, and $\circ$ the operation of function composition. Usual set-theoretic notation is employed throughout, for example, $\cup$, $\cap$, $-$, and $\in$ are respectively, set-theoretic union, intersection, difference, and membership. With a specific set X under consideration, $-$ is also considered as a complementation operator for subsets A of X, that is, $-A$ is taken to be $X - A$. $\subseteq$ is the subset relation, and $\subset$ is the proper subset relation, $\varnothing$ is the empty set, and $\wp(A)$ is the power set of A, $\{B | B \subseteq A\}$. $\notin$ stands for "is not a member of" and $\not\subseteq$ for "is not a subset of." For nonempty sets $\mathcal{E}$, $\bigcup \mathcal{E}$ and $\bigcap \mathcal{E}$ have the following definitions:

$$\bigcup \mathcal{E} = \{x | x \in E \text{ for some } E \text{ in } \mathcal{E}\} \text{ and } \bigcap \mathcal{E} = \{x | x \in E \text{ for all } E \text{ in } \mathcal{E}\}.$$

"iff" stands for "if and only if." $\quad\square$

The event spaces of standard probability theory are boolean algebras of events:

Definition 1.1 (boolean algebra of events). $\mathfrak{X} = \langle \mathcal{X}, \cup, \cap, -, X, \varnothing \rangle$ is said to be a *boolean algebra of events* if and only if X is a nonempty set, $-$ is the complementation operator with respect to X, and $\mathcal{X}$ is a set of subsets of X such that

- X, and $\varnothing$ are in $\mathcal{X}$,
- for all A and B in $\mathcal{X}$, $A \cup B$, $A \cap B$, and $-A$ are in $\mathcal{X}$.

Let $\mathfrak{X} = \langle \mathcal{X}, \cup, \cap, -, X, \varnothing \rangle$ be a boolean algebra of events. Elements of $\mathcal{X}$ are called *events (of $\mathfrak{X}$)*. X is called the *sure event (of $\mathfrak{X}$)* and $\varnothing$ *is called the null event (of $\mathfrak{X}$)*. $\mathfrak{X}$ is said to be *σ-additive* (or *countably additive*) if and only if for all denumerable sequences $A_1, \ldots, A_i, \ldots$ of $\mathcal{X}$, $\bigcup_{i=1}^{\infty} A_i$ is in $\mathcal{X}$. $\quad\square$

Definition 1.2 (traditional probability function). $\mathbb{P}$ is said to be a *traditional probability function* if and only if $\mathbb{P}$ is a function from a boolean algebra of events $\mathfrak{X} = \langle \mathcal{X}, \cup, \cap, -, X, \varnothing \rangle$ into the closed unit interval $[0, 1]$ of the reals such that

- $\mathbb{P}(\varnothing) = 0$ and $\mathbb{P}(X) = 1$, and
- *finite additivity:* for all A and B in $\mathcal{X}$, if $A \cap B = \varnothing$, then $\mathbb{P}(A \cup B) = \mathbb{P}(A) + \mathbb{P}(B)$.

If, in addition, $\mathfrak{X}$ is σ-additive and for each pairwise disjoint sequence of disjoint sets in $\mathcal{X}$, $A_1, \ldots, A_i, \ldots$, $\mathbb{P}(\bigcup_{i=1}^{\infty} A_i) = \sum_{i=1}^{\infty} \mathbb{P}(A_i)$, then $\mathbb{P}$ is said to be σ-additive (or *countably additive*). □

σ-additivity has an important role in many applications of probability theory. In many other applications only finite additivity is needed. In many situations, σ-additivity of a probability function has little theoretical or intuitive justification. In particular, foundations based on the convergence of relative frequencies of occurrence only give rise to finite additivity, and various axiomatic theories involving subjective probabilities only yield finite additivity. For the purposes of this book, the focus is on finite additivity.

The definition of "traditional probability function" and "σ-additive traditional probability function" in Definition 1.2 matches the definitions of "finitely additive measure" and "σ-additive measure" in analysis, with the measure of X, the universe under consideration, being 1. Despite this formal match, measure theory and probability theory are about different subject matters—measure theory about an abstract concept of "volume" and probability theory about the measurement of uncertainty. Because of this, different ideas are emphasized in the two sub-disciplines. In probability theory, the concepts of "conditional probability" and "independence" are of paramount importance.

Definition 1.3 (traditional conditional probability). Let $\mathbb{P}$ be a traditional probability function on the boolean algebra $\mathfrak{X} = \langle \mathcal{X}, \cup, \cap, -, X, \varnothing \rangle$ and A and B be elements of $\mathcal{X}$, where $\mathbb{P}(B) \neq 0$. Then the *traditional $\mathbb{P}$-conditional probability of A given B*, $\mathbb{P}(A \mid B)$, is, by definition,

$$\mathbb{P}(A \mid B) = \frac{\mathbb{P}(A \cap B)}{\mathbb{P}(B)}. \quad \square$$

Definition 1.4 (traditional independence). Let $\mathbb{P}$ be a probability function on the boolean algebra $\mathfrak{X} = \langle \mathcal{X}, \cup, \cap, -, X, \varnothing \rangle$ and A and B be elements of $\mathcal{X}$. Then A and B are said to be *traditionally $\mathbb{P}$-independent*, in symbols, $A \perp_{\mathbb{P}} B$, if and only if

$A = \varnothing$ or $B = \varnothing$ or $[A \neq \varnothing$ and $B \neq \varnothing$ and $\mathbb{P}(A \mid B) = \mathbb{P}(A)]$. □

Note that it follows from the definitions of "traditional $\mathbb{P}$-conditional probability" and "traditional $\mathbb{P}$-independence" that $A \perp_{\mathbb{P}} B$ if and only if $\mathbb{P}(A \cap B) = \mathbb{P}(A) \cdot \mathbb{P}(B)$.

As noted previously, while the concepts of "volume" from analysis and "uncertainty" from probability share many algebraic properties, ideas that are of importance for one may not be important for the other. For example, for analysis involving volumes it is not important to distinguish among sets of volume 0; for some issues involving uncertainty it is. As discussed earlier, the empty event, $\varnothing$, which has probability 0, is impossible: it cannot occur. However, in many situations there are events of probability 0 that may occur. Such events are improbable but not impossible. A comprehensive theory of probability requires that their occurrence be considered, that is, a comprehensive probability theory should account for properties of $\mathbb{P}(A \mid B)$, where $B \neq \varnothing$ and $\mathbb{P}(A) = 0$. In order to accomplish this, the definition of "traditional probability function" given in Definition 1.2 needs to be extended.

In this book, this is done by extending the definition of "traditional probability function" to a numerical system that includes not only the real number system, but also positive infinitesimal numbers. This extension, which is described in detail in Chapter 4, has all the relevant algebraic properties needed to define probabilities and carry out algebraic calculations in manners identical to their traditional counterpart that employs only the real number system. The basic idea is that a positive infinitesimal (a positive number that is $<$ each positive real number) is added to the real numbers, $\mathbb{R}$, to be part of a larger set of numbers ${}^\star\mathbb{R}$ that is the domain of a number system, ${}^\star\mathfrak{R}$. ${}^\star\mathfrak{R}$ is defined as follows.

An algebraic system ${}^\star\mathfrak{R} = \langle {}^\star\mathbb{R}, \leq, +, \cdot, 0, 1 \rangle$ is constructed for which $\mathbb{R} \subset {}^\star\mathbb{R}$, $\leq$ is a total ordering on ${}^\star\mathbb{R}$, and $+$ and $\cdot$ are binary operations on ${}^\star\mathbb{R}$ having the roles of, respectively, addition and multiplication, making ${}^\star\mathfrak{R}$ a totally ordered field, that is, an algebraic system that has all the ordinary algebraic properties involving the usual addition, multiplication, and ordering properties of the real number system. As shown in Chapter 4, it follows from these assumptions that the sum of two positive infinitesimals is a positive infinitesimal, and the product of a positive infinitesimal with a positive element β of $\mathbb{R}$ is a positive infinitesimal, and every element of

$$ {}^\star[0, 1] = \{x \mid 0 \leq x \leq 1\} $$

in ${}^\star\mathfrak{R}$ has the form $r + \alpha$, where r is a real number in the real interval $[0,1]$ and α is 0 or is a positive infinitesimal. The following definition extends the concept of "traditional probability function" to ${}^\star\mathfrak{R}$.

Definition 1.5 (extended probability function). $\mathbb{P}$ is said to be an *extended probability function* if and only if $\mathbb{P}$ satisfies the definition of "probability function" in Definition 1.2 with the following change and addition:

- $\mathbb{P}$ is a function from $\mathcal{X}$ into $^\star[0,1]$ (instead of into $[0,1]$), and
- *monotonicity:* for all A and B in $\mathcal{X}$, if A is a proper subset of B, then $\mathbb{P}(A) < \mathbb{P}(B)$.

In the obvious manner, the definitions of *extended conditional probability* and *extended independence* hold. □

The definitions of "traditional probability function" and "extended probability function" are formulated in terms of $\cup$, $\cap$, and $\subseteq$. Properties of the complementation operator, $-$, were not used. However, in order for the definitions of "traditional probability function" and "extended probability function" to be effective, there must be a reasonable supply of disjoint events. The assumption of a boolean algebra of events guarantees this through the existence of its complementation operator, $-$. Without such a supply of disjoint events, the condition,

$$\text{If } A \cap B = \varnothing, \text{ then } \mathbb{P}(A \cup B) = \mathbb{P}(A) + \mathbb{P}(B) ,$$

is of little use, and probability theory can degenerate into essentially a numerical ordering of events. For reasons that will become apparent, for purposes of generalization, it is desirable to have concepts of "probability function" whose effectiveness does not depend on the existence of a rich supply of disjoint events. This is made possible by using the following elementary result of probability theory.

Theorem 1.1. *Suppose $\mathbb{P}$ is a traditional probability function (or an extended probability function) on the boolean algebra $\mathfrak{X} = \langle \mathcal{X}, \cup, \cap, -, X, \varnothing \rangle$. Then for all A and B in $\mathcal{X}$,*

$$\mathbb{P}(A \cup B) = \mathbb{P}(A) + \mathbb{P}(B) - \mathbb{P}(A \cap B) . \tag{1.5}$$

Proof. Because $A = [A-(A\cap B)]\cup[A\cap B]$ and $[A-(A\cap B)]\cap[A\cap B] = \varnothing$, it follows from finite additivity (Definition 1.2) that

$$\mathbb{P}[A - (A \cap B)] = \mathbb{P}(A) - \mathbb{P}(A \cap B) .$$

Then, because $[A - (A \cap B)] \cap B = \varnothing$ and $[A - (A \cap B)] \cup B = A \cup B$,

$$\mathbb{P}(A) - \mathbb{P}(A \cap B) + \mathbb{P}(B) = \mathbb{P}(A \cup B) . \quad \square$$

Obviously, Equation 1.5 implies

$$\text{if } A \cap B = \varnothing, \text{ then } \mathbb{P}(A \cup B) = \mathbb{P}(A) + \mathbb{P}(B) , \tag{1.6}$$

and thus, by Theorem 1.1, Equation 1.5 can be used in place of Equation 1.6 in the definitions of "traditional probability function" and "extended probability function". With such a substitution, the definitions of "traditional probability function" and "extended probability function" applies effectively to situations with few or even no nontrivial disjoint events. This book makes use of this by using Equation 1.5 instead of Equation 1.6 in a definition of "probability function". As results in later chapters show, this provides for a richer and more general theory of probability than extended probability theory.

1.7 Lattices

Many kinds of structures have the same abstract algebraic properties of boolean algebras. Such structures are called "boolean lattices". They are special cases of more general class of algebras called "lattice algebras." Because lattice algebras are the only kind of lattices referred to in this book, they will often be called simply "lattices" in order to shorten definitions and statements. Functions with similar algebraic properties of extended probability functions can be defined on lattices. They are called "lattice probability functions".

Definition 1.6 (lattice). $\mathfrak{L} = \langle L, \sqcup, \sqcap, 1, 0 \rangle$ is said to be a *lattice algebra*, or just *lattice* for short, if and only if L is a nonempty set, $\sqcup$ and $\sqcap$ are binary operations on L, 1, called the *unit element*, is in L, 0, called the *zero element*, is in L, and the following eight axioms hold for all a, b, and c in L:

(1) $1 \sqcap a = a$ and $1 \sqcup a = 1$.
(2) $a \sqcup (b \sqcup c) = (a \sqcup b) \sqcup c$.
(3) $a \sqcup b = b \sqcup a$.
(4) $a \sqcup (a \sqcap b) = a$.
(5) $0 \sqcup a = a$ and $0 \sqcap a = 0$.
(6) $a \sqcap (b \sqcap c) = (a \sqcap b) \sqcap c$.
(7) $a \sqcap b = b \sqcap a$.
(8) $a \sqcap (a \sqcup b) = a$. □

Definition 1.6 follows a convention of lattice theory that uses 1 and 0 as special elements of a lattice, and as such they are not necessarily the real numbers 1 and 0. Sometimes other symbols will be used in place of 1 and

0 to denote the unit and zero elements.

The set algebra $\langle \mathcal{X}, \cup, \cap, X, \varnothing \rangle$ satisfies the definition of "lattice", where X is a nonempty set and $\mathcal{X}$ is a set of subsets of X such that X and $\varnothing$ are in $\mathcal{X}$, with $X = 1$, $\varnothing = 0$, $\cup = \sqcup$, and $\cap = \sqcap$.

Many important structures in mathematics satisfy the conditions of a lattice. In the literature, lattices have been viewed as generalizations of classical logic (e.g., Birkhoff & von Neumann, 1936). In this book, they are viewed as generalizations of domains of traditional probability functions, that is, used as domains for functions that measure uncertainty. The following theorem of Chapter 2 shows that each lattice is isomorphic to a special kind of lattice based on sets.

Theorem 1.2. *Let $\mathfrak{L} = \langle L, \sqcup, \sqcap, 1, 0 \rangle$ be a lattice. Then $\mathfrak{L}$ is isomorphic to a set lattice of the form $\langle \mathcal{X}, \uplus, \cap, X, \varnothing \rangle$, where X is a nonempty set and $\mathcal{X}$ is a set of subsets of X such that X and $\varnothing$ are in $\mathcal{X}$ and $\uplus$ is a binary operation on $\mathcal{X}$.*

Note that the isomorphism in Theorem 1.2 represents $\sqcap$ as set-theoretic intersection, $\cap$, but does not necessarily represent $\sqcup$ as set-theoretic union, $\cup$.

Because the set-theoretic subset relation, $\subseteq$, has a definition in terms of $\cap$ by,

$$A \subseteq B \ \text{ iff } \ A \cap B = A,$$

by the isomorphism in Theorem 1.2, the definition,

$$a \leq b \ \text{ iff } \ a \sqcap b = a,$$

defines a transitive and reflexive relation $\leq$ on the lattice $\mathfrak{L}$. The isomorphism also shows that 1 and 0 are, respectively, the maximal and minimal elements of $\mathfrak{L}$ in terms of the $\leq$ ordering. A theorem of Chapter 2 shows that $a \sqcup b$ and $a \sqcap b$ for arbitrary a and b in L are, respectively, the sup and inf of a and b in terms of the $\leq$ ordering.

The following definition generalizes the notion of "extended probability function" to lattices.

Definition 1.7 (lattice probability function; probabilistic lattice). $\mathbb{P}$ is said to be a *lattice probability function* on the lattice $\mathfrak{L} = \langle L, \sqcup, \sqcap, 1, 0 \rangle$ if and only if $\mathbb{P}$ is a function from L into the extended real interval $^\star[0, 1]$ such that the following three statements hold for all a and b in L:

(1) $\mathbb{P}(1) = 1$ and $\mathbb{P}(0) = 0$.

(2) *lattice monotonicity:* if $a < b$ then $\mathbb{P}(a) < \mathbb{P}(b)$.
(3) *lattice additivity:* $\mathbb{P}(a) + \mathbb{P}(b) = \mathbb{P}(a \sqcup b) + \mathbb{P}(a \sqcap b)$.

A lattice $\mathfrak{L}$ is said to be *lattice probabilistic,* or *L-probabilistic* for short, if and only if there exists a lattice probability function on it. □

Note that the concept of "L-probabilistic lattice" in Definition 1.7 generalizes the traditional concept of "probability" in the following three ways:

(a) It generalizes the range of "probability function" to include positive infinitesimal values.
(b) It generalizes the domain of "probability function" from a boolean algebra of events to a general lattice.
(c) It generalizes finite additivity (Definition 1.2) to lattice additivity (Definition 1.7).

Quantum physics employs a different generalization of traditional probability theory that restricts finite additivity to certain pairs of lattice elements. (See Definition 1.17 below.) This book investigates all these forms of generalization and apply them to substantive issues in the behavioral sciences.

The following definition and theorem characterizes lattices that correspond to boolean algebras of events.

Definition 1.8 (complemented, distributive, and boolean lattices). Let $\mathfrak{L} = \langle L, \sqcup, \sqcap, 1, 0 \rangle$ be a lattice. Then the following definitions hold:

- Let a be an element of L. Then b in L is said to be a *complement* of a if and only if

$$a \sqcap b = 0 \ \text{ and } \ a \sqcup b = 1,$$

  and $\mathfrak{L}$ is said to be *complemented* if and only if each element in L has a complement.
- Let $\mathfrak{L}$ be a complemented lattice. Then it follows from a use of the Axiom of Choice of set theory that there exists a function on L that assigns to each element a of L a complement of a. Such a function is called a *complementation operator* or *complementation operation* and usually be denoted by the symbol $-$ and less occasionally by the symbols $\neg$ and $^\perp$. Some lattices have many complementation operations.
- By the phrase "$\mathfrak{L} = \langle L, \sqcup, \sqcap, -, 1, 0 \rangle$ is a complemented lattice" is meant that $\mathfrak{L}$ is a complemented lattice and $-$ is a complementation operator on L.

- $\mathfrak{L}$ is said to be *distributive* if and only if for all a, b, and c in L,

$$a \sqcap (b \sqcup c) = (a \sqcap b) \sqcup (a \sqcap c).$$

- $\mathfrak{L}$ is said to be a *boolean lattice* if and only if it is complemented and distributive.   □

Theorem 2.17 of Chapter 2 shows that the assumption of a boolean lattice strengthens Theorem 1.2 to the following theorem.

Theorem 1.3 (Stone's representation theorem for boolean lattices). *Each boolean lattice is isomorphic to a boolean algebra of events.*

Theorem 1.3 combined with a theorem by Nikodym (1960) yield the following result.

Theorem 1.4. *Each boolean lattice has an extended probability function.*

A main goal of this book is to generalize traditional probability theory by generalizing its event space while retaining the rich algebraic structure implied by its probability function. With the event space is restricted to situations that have the very basic logical structure of a lattice, Stone's representation theorem suggests that one should look at lattices that generalize the concept of "complemented and distributive". Theorem 1.4 shows that the concept of "L-probabilistic" adds nothing more to lattices beyond what is achieved by the concept "boolean", that is, achieved jointly by the concepts of "complemented" and "distributive". Thus one reasonable strategy for seeking generalizations of traditional probability theory is to generalize "complemented" or generalize "distributive" in ways that "L-probabilistic" still applies to the lattice. This is one approach that is taken in this book.

Another reasonable strategy is to look for other lattice generalizations of "probability function". This is done in Chapter 6, where finite additivity is applied to only elements of the lattice that are "orthogonal" to one another. Such probability functions are called "orthoprobability functions" and the lattices on which they are defined are called "ortho-probabilistic". For ortho-probabilistic lattices to be non-boolean, distributivity needs to be generalized.

1.8   Forms of Complementation

Complemented and L-probabilistic are by themselves not sufficient conditions for obtaining boolean: Consider the case of a vector space of finite dimension $n > 1$. Its subspaces form a lattice $\mathfrak{X} = \langle \mathcal{X}, \uplus, \cap, X, 0 \rangle$ with X being the vector space, 0 being the null subspace of X consisting of the zero vector, $\mathcal{X}$ being the set of subspaces of X, $A \uplus B$ for A and B in $\mathcal{X}$ being the subspace generated by the vectors in $A \cup B$ (which is larger than $A \cup B$ when $A \neq 0$ and $B \neq 0$). This lattice is complemented: If $B_1, \ldots, B_n$ is a basis β for X, then a complement of a nontrivial subspace A is the subspace generated by $\beta - \alpha$, where α is the subset of basis elements in β that generate the subspace A. This complement is not the only complement of A in the lattice if the dimension of X is greater than 1. (As an example, let $\mathfrak{X}$ be a vector space with real scalars of dimension 2 and A and B be different 1-dimensional subspaces of X. Then $A \cap B = 0$ and $A \uplus B = X$. Let C be a 1-dimensional subspace of X such that $C \neq A$ and $C \neq B$. Then $A \cap C = 0$ and $A \uplus C = X$, that is, both B and C are complements of A.) Define $\mathbb{P}$ on $\mathcal{X}$ as follows: For each A in $\mathcal{X}$,

$$\mathbb{P}(A) = \frac{\dim(A)}{\dim(X)} = \frac{\dim(A)}{n},$$

where dim is the function that assigns each subspace of X its dimension. Then $\mathbb{P}$ is an extended lattice probability function on $\mathfrak{X}$, because it is a law of finite dimensional vector spaces that for all subspaces A and B in $\mathcal{X}$,

$$\dim(A) + \dim(B) = \dim(A \uplus B) + \dim(A \cap B).$$

$\mathfrak{X}$ is not boolean, because if $C = \{rc \,|\, r \in \mathbb{R}\}$ and $D = \{rd \,|\, r \in \mathbb{R}\}$, where c and d are linearly independent vectors in X (and thus C and D are 1-dimensional subspaces of X), then the 1-dimensional subspace $(C, D) = \{r(c + d) \,|\, r \in \mathbb{R}\}$ has the property,

$$(C, D) = (C, D) \cap (C \uplus D) \neq 0 \quad \text{and} \quad [(C, D) \cap C] \uplus [(C, D) \cap D] = 0,$$

contradicting distributivity, which is a key property of "boolean".

The complementation operation in a boolean algebra of events has special properties that do not generalize to all complemented lattices. Two of these are *De Morgan's Laws* and *unique complementation*. De Morgan Laws extend to an important class of non-boolean lattices for which probability theories have been developed. Unique complementation extends to some non-boolean lattices, but such lattices apparently do not have interesting mathematical properties outside of their existence, and have not appeared in science or probabilistic theories.

Definition 1.9 (De Morgan's Laws). Suppose $\mathfrak{L} = \langle L, \sqcup, \sqcap, -, 1, 0 \rangle$ is a complemented lattice and a and b are in L. Then *De Morgan's Laws* are satisfied if and only if

(i) $-(a \sqcup b) = (-a) \sqcap (-b)$, and
(ii) $-(a \sqcap b) = (-a) \sqcup (-b)$. $\square$

It is well known that boolean algebras events satisfy De Morgan's Laws (e.g., see Theorem 2.15 of Chapter 2). Thus by Stone's Representation Theorem, all boolean lattices satisfy De Morgan's Laws. Some non-boolean lattices also satisfy De Morgan's Laws. For example, the 2-dimensional real vector space $\mathfrak{X}$ described above was shown to be non-boolean. When for each subspace A of $\mathfrak{X}$, $A^{\perp}$ is defined as the orthogonal complement A, it is a well-known result of vector space theory that $^{\perp}$ is a complementation operation satisfying De Morgan's Laws.

The following theorem provides another well-known property of boolean algebras that by Stone's Representation Theorem holds for general boolean algebras.

Definition 1.10 (unique complementation). A lattice

$$\mathfrak{L} = \langle L, \sqcup, \sqcap, 1, 0 \rangle$$

is said to be *uniquely complemented* if and only if it is complemented and for all complementation operations $-$ and $-'$ on $\mathfrak{L}$, $- = -'$. $\square$

Theorem 1.5. *Each boolean lattice is uniquely complemented.*
      Proof. Theorem 2.18 of Chapter 2. $\square$

The following is a different kind of "complementation" operator that plays an important role in this book:

Definition 1.11 (pseudo complement). Let $\mathfrak{L} = \langle L, \sqcup, \sqcap, 1, 0 \rangle$ be a lattice and a be in L. Then b is said to be the *pseudo complement* of a if and only if $b \in L$, $b \sqcap a = 0$, and for all c in L, if $c \sqcap a = 0$, then $c \leq b$. $\mathfrak{L}$ is said to be *pseudo complemented* if and only if each element of L has a pseudo complement. $\square$

Theorem 1.6. *Let $\mathfrak{L} = \langle L, \sqcup, \sqcap, 1, 0 \rangle$ be a lattice and a be an element of L. Then the pseudo complement of a, if it exists, is unique.*
      Proof. Suppose b and c are pseudo complements of a. Then $b \sqcap a = 0$ and $c \sqcap a = 0$. Then, by the definition of "pseudo complement", $c \leq b$ and $b \leq c$, and thus, because $\leq$ is antisymmetric, $b = c$. $\square$

Definition 1.12 ($\bar{}$). By Theorem 1.6, the operation of pseudo complementation of a pseudo complemented lattice is unique. Unless otherwise stated, $\vdash$ will often be used to denote the operation of pseudo complementation of a pseudo complement lattice. $\square$

In logic and the foundations of mathematics the following concept, which implies pseudo complementation, plays an important role. However, this role is difficult to comprehend in the following definition. Chapter 3 develops additional concepts that allows for it to given a more intuitive formulation that makes clear its connection to logic.

Definition 1.13 (relative pseudo complementation, $\Rightarrow$). Let $\mathfrak{L} = \langle L, \sqcup, \sqcap, 1, 0 \rangle$ be a lattice and a and b be in L. Then the element $a \Rightarrow b$ of L is said to be the *pseudo complement of a relative to b* if and only if

- $a \sqcap (a \Rightarrow b) \leq b$, and
- for all x in L, if $a \sqcap x \leq b$, then $x \leq (a \Rightarrow b)$. $\square$

Definition 1.14 (heyting lattice). $\mathfrak{L} = \langle L, \sqcup, \sqcap, 1, 0 \rangle$ is said to be a *heyting lattice* if and only if it is a lattice and for all a and b in L, $a \Rightarrow b$ exists. By convention, heyting lattices will often be written as $\mathfrak{L} = \langle L, \sqcup, \sqcap, \Rightarrow, 1, 0 \rangle$. $\square$

The following theorem is immediate from Definitions 1.13 and 1.14.

Theorem 1.7. *Suppose $\mathfrak{L} = \langle L, \sqcup, \sqcap, \Rightarrow, 1, 0 \rangle$ is a heyting lattice. Then $\mathfrak{L}$ is pseudo complemented and for all a in L,*

$$\vdash a \ = \ a \Rightarrow 0 \,,$$

where $\vdash$ is the operation of pseudo complementation for $\mathfrak{L}$. $\square$

The following theorem is shown in Chapter 3 (Theorem 3.3).

Theorem 1.8. *Suppose $\mathfrak{L} = \langle L, \sqcup, \sqcap, \Rightarrow, 1, 0 \rangle$ is a heyting lattice. Then the following two statements hold.*

(1) $\mathfrak{L}$ is distributive.
(2) $\mathfrak{L}$ is L-probabilistic.

Theorems 1.7 and 1.8 above and Theorems 1.9 and 1.10 below suggest that a pseudo complemented distributive lattice is a good candidate for an event structure for a generalized theory of probability.

The following theorem is a consequence of Theorem 4.15 of Chapter 4.

Theorem 1.9. *Suppose $\mathfrak{L}$ is a pseudo complemented distributive lattice. Then $\mathfrak{L}$ is L-probabilistic.* $\square$

The following surprising theorem is shown in Chapter 2 (Theorem 2.21).

Theorem 1.10. *Let $\mathfrak{L}$ be a lattice. Then the following two statements are logically equivalent.*

(1) There exists an operation $\vdash$ that is both a complementation and a pseudo complementation operation of $\mathfrak{L}$.

(2) $\mathfrak{L}$ is boolean. □

Theorem 1.10 suggests the following two approaches for generalizing traditional probability theory:

(1) Base the theory on pseudo complemented lattices.
(2) Base the theory on complemented lattices that are not pseudo complemented.

In order to achieve interesting generalizations of traditional probability theory, both approaches need to be supplemented by additional assumptions. This book's approach to (1) is to assume a pseudo complemented distributive lattice. For (2) to yield an interesting generalization, the lattice must be non-distributive by Theorem 1.10. A fruitful approach to (2) that has been exploited in quantum mechanics and other applications is to generalize the concept of "probability function" so that it is additive only over lattice elements that are "orthogonal" to one another. There is a large, highly sophisticated literature on this topic, and the usual approaches to it are not pursued in this book. (The reader interested in this literature should consult Kalmbach, 1983.) Instead, in Chapter 6, a different foundational theory for "lattice orthogonality" is presented that I consider to be a better match for the concerns and methods of the behavioral sciences.

The following definitions and Theorem 1.11 below capture the key lattice concepts related to such "orthogonal elements".

Definition 1.15 (orthocomplement, $^\perp$; orthogonal). Let

$$\mathfrak{L} = \langle L, \sqcup, \sqcap, 1, 0 \rangle$$

be a lattice. Then $^\perp$ is said to be an *orthocomplement* of $\mathfrak{L}$ if and only if it is a complementation operation on $\mathfrak{L}$ that satisfies De Morgan's Laws.

By convention, "$\mathfrak{L} = \langle L, \sqcup, \sqcap, ^\perp, 1, 0 \rangle$ is an orthocomplemented lattice" stands for "$\mathfrak{L}$ is a lattice and $^\perp$ is an orthocomplement on $\mathfrak{L}$."

Let $\mathfrak{L} = \langle L, \sqcup, \sqcap, ^\perp, 1, 0 \rangle$ be an orthocomplemented lattice and a and b be elements of L. Then a and b are said to be *orthogonal*, in symbols, $a \perp b$, if and only if $b \leq a^{\perp}$. Note that this use of symbol "$\perp$" is different from the use of the symbol "$\perp$" for the concept of probabilistic independence. □

Definition 1.16 (ortholattice). An orthocomplemented lattice is called an *ortholattice*. □

Theorem 1.11. *Suppose* $\mathfrak{L} = \langle L, \sqcup, \sqcap, ^{\perp}, 1, 0 \rangle$ *is an ortholattice and that* $^{\perp}$ *is the only complementation operation for* $\mathfrak{L}$. *Then* $\mathfrak{L}$ *is boolean.*
  Proof. Theorem 2.19 of Chapter 2. □

Theorem 1.11 implies that ortholattices that are proper generalizations of "boolean". When they are non-boolean, they are not only non-distributive but also must have multiple complementation operations.

Definition 1.17 (orthoprobability function; orthoprobabilistic lattice). Let $\mathfrak{L} = \langle L, \sqcup, \sqcap, ^{\perp}, 1, 0 \rangle$ be an ortholattice. Then $\mathbb{P}$ is said to be an *orthoprobability function* on $\mathfrak{L} = \langle L, \sqcup, \sqcap, 1, 0 \rangle$ if and only if $\mathbb{P}$ is a function from L into the extended real interval $^{\star}[0,1]$ such that for all a and b in L,

- $\mathbb{P}(1) = 1$ and $\mathbb{P}(0) = 0$,
- *lattice monotonicity:* If $a < b$ then $\mathbb{P}(a) < \mathbb{P}(b)$,
- *orthoadditivity:* If $a \perp b$ then $\mathbb{P}(a) + \mathbb{P}(b) = \mathbb{P}(a \sqcup b)$.

An ortholattice $\mathfrak{L}$ is said to be *orthoprobabilistic* if and only if there exists an orthoprobability function on it. □

Definition 1.18 (orthomodular). $\mathfrak{L} = \langle L, \sqcup, \sqcap, 1, 0 \rangle$ is said to be *orthomodular* if and only if it is an ortholattice and for all a and b in L,

$$\text{if } a \leq b \text{ then } a \sqcup (a^{\perp} \sqcap b) = b. \tag{1.7}$$

Equation 1.7 is called the *orthomodular law*. □

Theorem 2.24 of Chapter 2 shows that distributive lattices are orthomodular. Theorem 6.1 of Chapter 6 shows the following useful result.

Theorem 1.12. *Suppose the ortholattice* $\mathfrak{L}$ *is orthoprobabilistic. Then* $\mathfrak{L}$ *is orthomodular.*

Orthomodular lattices and their associated logics and probability theories have a long history and a well-developed literature, usually under the rubric "quantum logic". They, along with probability theories based on pseudo complementation and heyting lattices, are applicable to a variety of experimental and theoretical phenomena. In particular, orthomodular and pseudo complemented lattices can model and justify various kinds of puzzling decision phenomena, some of which are considered as "irrational" by economists, philosophers and others, and Chapter 6 shows that orthomodular lattices can provided new insights to how psychological experiments are interrelated.

1.9   Rationality

Various arguments for the rationality of traditional probability theory have been put forth in the literature. One of the main ones is the Dutch Book argument. It assumes a boolean algebra of events $\mathcal{X}$, for which an individual, called a *bookie,* sets prices for lotteries having the following form: For each event A in $\mathcal{X}$,

$$\text{pay \$1 if } A \text{ occurs and pay \$0 if } A \text{ does not occur.}$$

The bookie is committed to buying or selling any number of such lotteries. A *Dutch Book* said to be able to made against the bookie if her prices can be arbitraged—that is, if an individual, called *an arbitrageur* can buy lotteries from her and sell lotteries to her at the prices in a manner so that he will always make a profit, no matter what event occurs. The bookie prices for the lotteries is said to be *(Dutch Book) rational* if no Dutch Book can be made against the bookie. The *Dutch Book Theorem* (Section 4.2 in Chapter 4) states that no Dutch Book can be made against the bookie if and only if the prices on events from $\mathcal{X}$ have the properties of a traditional probability function.

In Chapter 4 the Dutch Book Theorem is extended to situations where the event space is a distributive set lattice. Let $\mathfrak{S} = \langle \mathcal{E}, \cup, \cap, X, \varnothing \rangle$ be a distributive set lattice. For $\mathfrak{S}$, an Arbitrageur might have fewer ways to formulate buying and selling opportunities than if $\mathcal{E}$ were extended to boolean algebra of events. This might cause worry about whether "no Dutch Book can be made" is a proper test for rationality for $\mathfrak{S}$. However, a theorem in Chapter 4 shows that for $\mathfrak{S}$, no Dutch Book can be made against the Bookie if and only if the Bookie's prices for the events in $\mathcal{E}$ can be extended to a finitely additive measure $\mathbb{P}$ on boolean algebra of events $\mathfrak{B}$ with domain the power-set of X. $\mathfrak{B}$ obviously has maximum richness for a boolean algebra for formulating buying and selling opportunities. This implies that the Bookie's prices on events $\mathcal{S}$ are rational, if "rationality" is to be defined according to the usual formulation of the Dutch Book Theorem.

Another widely used approach to rationality is the subjective expected utility model (SEU), which is used throughout science as *the* model for rational decision making. It employs a finitely additive probability function on a boolean algebra of events for measuring uncertainty. A few have argued that SEU is too narrow for a general, rational model of decision making. Chapter 5 explores more general alternatives to SEU based on pseudo complemented distributive lattices and rationality considerations. One of these alternatives provides a foundation for an influential descriptive psychological theory of probability estimation known as "support theory". Another emphasizes a form of "bounded rationality" that is applicable to situations where the decision maker enters into various states while evaluating lotteries.

1.10   Logic

Besides being a formal generalization of the domain of a probability function, that is, a generalization of a boolean algebra of events, a lattice $\mathfrak{L} = \langle L, \sqcup, \sqcap, 1, 0 \rangle$ can also be viewed as a formal generalization of classical logic. In such generalizations, L is interpreted as a set of propositions, $\sqcup$ as a disjunction operator (i.e., a generalization of "or"), $\sqcap$ as a conjunction operator (i.e., a generalization of "and"), 1 as a tautology, and 0 as a contradiction. Because $\sqcup$ and $\sqcap$ are operators on L, formulas built up out of propositions in L using the operations $\sqcup$ and $\sqcap$ are propositions of $\mathfrak{L}$, for example, $a \sqcap [(b \sqcup c)] \sqcap 1$ is a proposition of $\mathfrak{L}$ for a, b, and c in L. $\alpha = \beta$, for propositions α and β of $\mathfrak{L}$, are interpreted as "α and β are logically equivalent". Thus the proposition, $a \sqcup b = 1$, says in lattice logic language that the proposition $a \sqcup b$ is logically equivalent to the tautology 1. Alternatively, this is the same as saying that "$a \sqcup b$ is a tautology" or saying that "$a \sqcup b$ is logically true." Similarly, and the proposition $a \sqcap b = 0$ says in lattice logic language that the proposition $a \sqcap b$ is a contradiction, which is the same as saying "$a \sqcap b$ is logically false."

In the literature, lattices have two principal types of interpretations: as logics and as event spaces. These two types of interpretations are about different subject matters and should not be confused. Logics are primarily concerned with the notion of "logical consequence," and probability theory with "degrees of belief," or "propensity to occur."

A logical implication relation plays a major role in logic, because of logic's emphasis on deduction and logical consequence. In probability theory, $a \leq b$ corresponds to the idea that b is a consequence of a, and this gets expressed probabilistically by the conditional probability of b given a is 1. An analog of a theory of deduction of formal logic is not used in probabilistic scientific applications. Because of this, this book does not undertake a development of "lattices as logics."

Chapter 2

Basic Lattice Theory

2.1 Introduction

For the purposes of readability and continuity, various concepts of Chapter 1 will be reintroduced in this or subsequent chapters, often with supplementary material or comments. Many theorems stated in Chapter 1 did not have proofs. These too will be restated and proofs or detailed references to them will be supplied.

Lattices are simple algebraic structures that have numerous interpretations. For the purposes of this book, they are used as domains for probability functions and for isolating algebraic structures inherent in logics. This chapter employs concepts and theorems from lattice theory to:

(i) provide a generalization of "probability function" that will be used in later chapters, and

(ii) provide a picture of the kinds of lattices satisfying the generalized probability concept.

The following are a few examples of lattices from mathematics. They illustrate some of the importance differences in algebraic properties among lattices.

Lattice Examples In the following examples $\mathfrak{X} = \langle \mathcal{X}, \Cup, \cap, X, \varnothing \rangle$, where X and $\varnothing$ are elements of $\mathcal{X}$.

(1) $\Cup = \cup$ and $\mathfrak{X}$ is a boolean algebra of subsets of X.

(2) $\Cup = \cup$ and $\mathcal{X}$ is a topology with universe X.

(3) $\mathcal{X}$ is the set of all normal subgroups of a group X and $A \Cup B$ is the smallest normal subgroup of X containing the normal subgroups A and B of X.

(4) $\mathcal{X}$ is the set of all subspaces of a finite dimensional vector space X and $A \uplus B$ is the smallest vector subspace of X containing the subspaces A and B of X.

(5) $\mathcal{X}$ is the set of all closed subspaces of a hilbert space X and $A \uplus B$ is the smallest closed subspace of H containing the subspaces A and B of X. $\quad\square$

Notice as a lattice, (1) is a boolean and (2) is heyting. Both are distributive. (3), (4), and (5) are not distributive, but can be shown to satisfy the *modular law*,

$$\text{If } A \subseteq B, \text{ then } A \cap (B \uplus C) = (A \cap B) \uplus (A \cap C),$$

for all A, B, and C in $\mathcal{X}$. Obviously, all distributive lattices are modular. Thus all the above lattices are modular. (1), (3), (4), and (5) have complementation operations, and (2) does not, except in very special situations. The complementation operation in (1) is unique (e.g., see Theorem 2.13 below). However, only in special circumstances are the complementation operations in (3), (4), and (5) unique. (2) does not have a complementation operation except for the special case when it is boolean. It does, however, have a unique pseudo complementation operation.

Distributivity, modularity, complementation, and pseudo complementation are important lattice properties of various kinds of functions that generalize the classical notion of "probability function". This chapter covers elementary results involving these properties. Chapters 3 delves more deeply into the case of a pseudo complemented and distributive lattice, and Chapters 4 and 5 provide philosophical and scientific applications of probability theory on pseudo complemented and distributive lattices. For some scientific applications of probability theory, modularity has to be further generalized to orthomodularity.

2.2   General Notation, Conventions, and Definitions

The following notation, conventions, and definitions involving orderings are employed throughout this book.

Definition 2.1 (denumerability). Let X be a set. Then X is said to be *denumerable* if and only if there exists a one-to-one function from the set of positive integers, $\mathbb{I}^+$, onto X. X is said to be *countable* if and only if it is denumerable or finite. $\quad\square$

Definition 2.2 (ordering properties). Let X be a nonempty set and $\precsim$ be a binary relation on X. Then $\precsim$ is said to be

- *reflexive* if and only if for all x in X, $x \precsim x$,
- *transitive* if and only if for all x, y, and z in X, if $x \precsim y$ and $y \precsim z$ then $x \precsim z$,
- *symmetric* if and only if for all x and y in X, if $x \precsim y$ then $y \precsim x$,
- *connected* if and only if for all x and y in X, either $x \precsim y$ or $y \precsim x$,
- *antisymmetric* if and only if for all x and y in X, if $x \precsim y$ and $y \precsim x$, then $x = y$.

Define the binary relations $\prec$, $\succsim$, $\succ$, and $\sim$ terms of $\precsim$ as follows: For all x and y in X,

- $x \prec y$ if and only if $x \precsim y$ and not $y \precsim x$,
- $x \succsim y$ if and only if $y \precsim x$,
- $x \succ y$ if and only if $y \prec x$,
- $x \sim y$ if and only if $x \precsim y$ and $y \precsim x$. □

Definition 2.3 (kinds of ordering relations). Let $\precsim$ be a binary relation on the nonempty set X. Then $\precsim$ is said to be

- a *partial ordering* on X if and only if $\precsim$ is a reflexive, transitive, and antisymmetric relation on X,
- a *weak ordering* if and only if $\precsim$ is transitive and connected,
- a *total ordering* if and only if $\precsim$ is a weak ordering and is antisymmetric.

Note that weak orderings are also partial orderings, so each concept in the list is a generalization of the next.

It is immediate that weak and total orderings are reflexive. Let $\precsim$ be a partial or total ordering. By convention, $\precsim$ is usually written as $\preceq$ to emphasize the fact that the relation $\sim$ defined in terms of $\precsim$ is the identity relation, $=$. □

Definition 2.4 (inf and sup). Let $\preceq$ be a partial ordering on the set X and a, b, and c be arbitrary elements of X. Then $c = \sup\{a, b\}$ (the *supremum of a and b)* if and only if

$$a \preceq c,\ b \preceq c,\ \text{and for all } e \text{ in } X, \text{ if } a \preceq e \text{ and } b \preceq e \text{ then } c \preceq e,$$

and $d = \inf\{a, b\}$ (the *infimum of a and b)* if and only if

$$d \preceq a,\ d \preceq b,\ \text{and for all } e \text{ in } X, \text{ if } e \preceq a \text{ and } e \preceq b \text{ then } e \preceq d. \quad □$$

The following lemma is immediate from Definition 2.4.

Lemma 2.1. *Suppose $\preceq$ is a partial ordering on the set X. Then the following two statements hold for all a and b in X.*

(1) If $\sup\{a, b\}$ exists then it is unique.
(2) If $\inf\{a, b\}$ exists then it is unique. $\square$

Definition 2.5 (equivalence relation). A relation $\equiv$ on a nonempty set is said to be an *equivalence relation* if and only if it is reflexive, transitive, and symmetric. $\square$

It follows that if $\precsim$ is a weak ordering on X, then $\sim$ is an equivalence relation on X.

The following definition is useful for distinguishing how the usual total ordering of the real numbers differs from the usual total ordering of the rational numbers.

Definition 2.6 (Dedekind cut and completeness). Suppose $\preceq$ is a total ordering on X. Then (A, B) is said to be a *Dedekind cut* of $\langle X, \preceq \rangle$ if and only if

- A and B are nonempty subsets of X,
- $A \cup B = X$,
- and for each x in A and each y in B, $x \prec y$.

Suppose (A, B) is a Dedekind cut of $\langle X, \preceq \rangle$, where $\preceq$ is a total ordering on X. Then c is said to be a *cut element* of (A, B) if and only if either

- c is in A and $x \preceq c \prec y$ for each x in A and each y in B,

or

- c is in B and $x \prec c \preceq y$ for each x in A and each y in B.

$\langle X, \preceq \rangle$ is said to be *Dedekind complete* if and only if each Dedekind cut of $\langle X, \preceq \rangle$ has a cut element. $\square$

The following theorem is well-known, and its proof is left to the reader.

Theorem 2.1. $\langle \mathbb{R}, \leq \rangle$ *is Dedekind complete, and for each Dedekind cut (A, B) of $\langle \mathbb{R}, \leq \rangle$, if r and s are cut elements of (A, B), then $r = s$.* $\square$

Definition 2.7 (continuum). $\langle X, \preceq \rangle$ is said to be a *continuum* if and only if the following four statements hold:

(1) *Total ordering*: $\preceq$ is a total ordering on X (Definition 2.3).
(2) *Unboundedness*: $\langle X, \preceq \rangle$ has no $\prec$-greatest or $\prec$-least element.
(3) *Denumerable density*: There exists a denumerable subset Y of X such that for each x and z in X, if $x \prec z$ then there exists y in Y such that $x \prec y$ and $y \prec z$.
(4) *Dedekind completeness*: $\langle X, \preceq \rangle$ is Dedekind complete. (Definition 2.6).

Theorem 2.2 (Cantor's Continuum Theorem). $\mathfrak{X} = \langle X, \preceq \rangle$ *is a continuum if and only if $\mathfrak{X}$ is isomorphic to $\langle \mathbb{R}^+, \leq \rangle$.*
    Proof. Cantor (1895). (A proof is also given in Theorem 2.2.2 of Narens, 1985.) $\square$

2.3   Lattices: Definitions and Duality

Lattices are simple algebraic structures that have many substantive interpretations. For example, in this book they are used as

- abstract algebraic structures,
- event spaces for various kinds of probability or belief functions, and
- models and truth tables for logics.

    There are two principal ways to define lattices. The first is the method used in Chapter 1, by which a lattice is defined as an algebra involving two operations, $\sqcup$ and $\sqcap$, a unit element, 1, and a zero element, 0, satisfying certain axioms. The second method defines a lattice as a partially ordered set with a maximal element 1 and a minimal element 0 satisfying certain axioms. This section begins with the first method involving two operations. The definition of "lattice" of Chapter 1 is repeated, but with its defining conditions given as named axioms. A description is then given of the "Duality Principle", which enables a particular method used in lattice theory for shortening proofs. Then the second formulation of "lattice" via a partial ordering is derived as a theorem from the first method's definition of "lattice".

Definition 2.8 (lattice). $\mathfrak{L} = \langle L, \sqcup, \sqcap, 1, 0 \rangle$ is said to be a *lattice algebra*, or just *lattice* for short, if and only if L is a nonempty set, $\sqcup$ and $\sqcap$ are binary operations on L, $1 \in L$, $0 \in L$, and the following eight axioms hold for all a, b, and c in L:

Axiom L1 [unit element] $1 \sqcap a = a$ and $1 \sqcup a = 1$. (1 is called the *unit* *element* of $\mathfrak{L}$.)

Axiom L2 [join associativity identity] $a \sqcup (b \sqcup c) = (a \sqcup b) \sqcup c$. ($a \sqcup b$ is called the *join* of a and b, and $\sqcup$ is called the *join operation* of L.)

Axiom L3 [join commutativity identity] $a \sqcup b = b \sqcup a$.

Axiom L4 [join absorption identity] $a \sqcup (a \sqcap b) = a$.

Axiom L5 [zero element] $0 \sqcup a = a$ and $0 \sqcap a = 0$. (0 is called the *zero element* of $\mathfrak{L}$.)

Axiom L6 [meet associativity identity] $a \sqcap (b \sqcap c) = (a \sqcap b) \sqcap c$. ($a \sqcap b$ is called the *meet* of a and b, and $\sqcap$ is called the *meet operation* of L.)

Axiom L7 [meet commutativity identity] $a \sqcap b = b \sqcap a$.

Axiom L8 [meet absorption identity] $a \sqcap (a \sqcup b) = a$.   □

The axioms L1 to L8 are formulated symmetrically in terms of $\sqcup$ and $\sqcap$ and 1 and 0. This symmetry will be used to shorten certain proofs through the use of the "Duality Principle" (Theorem 2.3 below).

Definition 2.9 (lattice theoretical proposition, dual). Let

$$\mathfrak{L} = \langle L, \sqcup, \sqcap, 1, 0 \rangle$$

be a lattice. A *lattice theoretical proposition* is a first order statement Θ involving $\sqcup$, $\sqcap$, 1, and 0, $=$, the logical connectives "or", "and", "not", "if ... then", "if and only if", the quantifiers (applied to elements of L) "for all" and "for some", or variants or combinations of the preceding, for example, "there exists" or "neither ... nor".

The *dual of* Θ is the statement resulting from Θ by the following substitutions:

$$\sqcup \mapsto \sqcap, \quad \sqcap \mapsto \sqcup, \quad 1 \mapsto 0, \quad 0 \mapsto 1. \quad \square$$

It immediately follows that the dual of the dual of Θ is Θ.

In the axiomatization of "lattice" given in Definition 2.8, Axioms L5 to L8 are duals of Axioms L1 to L4 and vice versa. Because in Definition 2.8 the dual of each lattice axiom is another lattice axiom, the following principle holds:

Theorem 2.3 (Duality Principle). *The dual of any lattice theoretical proposition derived from the lattice axioms is also derivable from the lattice axioms.*

Proof. A derivation of a lattice theoretical proposition Θ consists of a sequence $\Lambda_1, \ldots, \Lambda_n, \Theta$ of lattice theoretical propositions. Then the following is a method for deriving the dual of Θ: Produce the sequence $\mathcal{D}(\Lambda_1), \ldots, \mathcal{D}(\Lambda_n), \mathcal{D}(\Theta)$, where $\mathcal{D}(\Lambda)$ is the dual of Λ. □

The Duality Principle's main use is for shortening proofs. It does not add any deep insight into discovering proofs.

The following theorem's proof makes a simple use of the Duality Principle.

Theorem 2.4. *Suppose $\mathfrak{L} = \langle L, \sqcup, \sqcap, 1, 0 \rangle$ is a lattice. Then the following two statements hold.*

(1) Join idempotence: *For all a in L, $a \sqcup a = a$.*
(2) Meet idempotence: *For all a in L, $a \sqcap a = a$.*

Proof. 1. Let a and b be arbitrary elements of L and $c = a \sqcap b$. By join absorption,

$$a \sqcap a = a \sqcap [a \sqcup (a \sqcap b)] = a \sqcap (a \sqcup c).$$

By meet absorption,

$$a \sqcap (a \sqcup c) = a.$$

Thus, $a \sqcap a = a$, and join idempotence holds.

2. By duality, meet idempotence holds. □

The following theorem includes another application of the Duality Theorem.

Theorem 2.5. *Suppose $\mathfrak{L} = \langle L, \sqcup, \sqcap, 1, 0 \rangle$ is a lattice. Then the following two statements hold.*

(1) *For all a and b in $\mathcal{L}$, if $a \sqcap b = a$, then $a \sqcup b = b$.*
(2) *For all a and b in $\mathcal{L}$, if $a \sqcup b = a$, then $a \sqcap b = b$.*

Proof. (1). Let a and b be arbitrary elements of L such that $a \sqcap b = a$. Then

$$(a \sqcap b) \sqcup b = a \sqcup b.$$

By join commutativity and join absorption,

$$(a \sqcap b) \sqcup b = b \sqcup (a \sqcap b) = b$$

and thus $a \sqcup b = b$.

(2). Because Statement (1) holds and Statement (2) is the dual of Statement (1), Statement (2) holds by the Duality Principle.   □

Definition 2.10 ($\leq$,$\nleq$). Suppose $\mathfrak{L} = \langle L, \sqcup, \sqcap, 1, 0 \rangle$ is a lattice. $\leq$ denotes the binary relation on L such that for all a and b in L,

$$a \leq b \text{ iff } a \sqcap b = a \,,$$

and $\nleq$ denotes the binary relation on L such that for all a and b in L,

$$a \nleq b \text{ iff it is not the case that } a \leq b. \quad □$$

The notation "$\leq$", besides signifying a relation on general lattices, is also used to denote the standard ordering on the reals and nonempty subsets of the reals. It is usually clear from context which of these interpretations of $\leq$ is intended. When not, a clarifying phrase like "the lattice ordering $\leq$" will be added.

Extensions of the Duality Principle. By Definition 2.10, $a \leq b$ iff $a \sqcap b = a$. Thus the dual of $a \leq b$ is $a \sqcup b = a$, which by Statement (2) of Theorem 2.5 is alternatively formulated as $b \leq a$. Because of this, the Duality Principle is extended to include $\leq$ as follows: If an expression of the form $\alpha \leq \beta$ occurs in a lattice theoretical proposition Θ, then along with the other substitutions in computing the dual of Θ, the following substitution is made:

$$\alpha \leq \beta \mapsto \beta \leq \alpha.$$

It is easy to see that the Duality Principle also extends to include expressions of the form $\inf \{\alpha, \beta\}$ and $\sup \{\alpha, \beta\}$, because of the way they are defined in terms of $\leq$: Along with the previous substitutions in computing the dual of Θ, the following substitutions are made:

$$\inf \{\alpha, \beta\} \mapsto \sup \{\alpha, \beta\} \text{ and } \sup \{\alpha, \beta\} \mapsto \inf \{\alpha, \beta\}. \quad □$$

Theorem 2.6. *Suppose* $\mathfrak{L} = \langle L, \sqcup, \sqcap, 1, 0 \rangle$ *is a lattice. Then the following two statements hold.*

(1) $\leq$ *(Definition 2.10) is a partial order on* L.
(2) For all a *and* b *in* L, $a \sqcap b = \inf \{a, b\}$ *and* $a \sqcup b = \sup \{a, b\}$.

Proof. 1. Let a, b, and c be arbitrary elements of L.

(1). It is immediate from Definition 2.10 that $\leq$ is reflexive, that is $a \leq a$.

To show $\leq$ is antisymmetric, suppose $a \leq b$ and $b \leq a$. Then by Definition 2.10, $a \sqcap b = a$ and $b \sqcap a = b$. Thus, because by the commutativity of $\sqcap$, $b \sqcap a = a \sqcap b$, it follows that $a = b$.

To show $\leq$ is transitive, suppose $a \leq b$ and $b \leq c$. Then

$$a \sqcap b = a \quad \text{and} \quad b \sqcap c = b.$$

Thus by $\sqcap$-associativity,

$$a \sqcap c = (a \sqcap b) \sqcap c = a \sqcap (b \sqcap c) = a \sqcap b = a,$$

and thus $a \leq c$.

(2). Using lattice theoretical identities, the following shows that $a \sqcap b$ is a lower bound for a and b:

$$(a \sqcap b) \sqcap a = a \sqcap (a \sqcap b) = (a \sqcap a) \sqcap b = a \sqcap b,$$

and thus by the definition of $\leq$, $(a \sqcap b) \leq a$. Similarly,

$$(a \sqcap b) \sqcap b = a \sqcap (b \sqcap b) = a \sqcap b,$$

and thus $(a \sqcap b) \leq b$.

It will now be shown that $a \sqcap b = \inf \{a, b\}$. Suppose d is another lower bound for a and b. Then $d \leq a$ and $d \leq b$, and thus by the definition of $\leq$,

$$d \sqcap a = d \quad \text{and} \quad d \sqcap b = d.$$

From this it follows that

$$d \sqcap (a \sqcap b) = (d \sqcap a) \sqcap b = d \sqcap b = d,$$

and thus $d \leq (a \sqcap b)$, establishing that $a \sqcap b = \inf \{a, b\}$.

Because $a \sqcap b = \inf \{a, b\}$, and because the dual of $a \sqcap b = \inf \{a, b\}$ is $a \sqcup b = \sup \{a, b\}$, it follows from the Duality Principle that $a \sqcup b = \sup \{a, b\}$. $\square$

Theorem 2.7. *Suppose $\preceq$ is a partial order on the set L, $1 \in L$ is the maximal element of L with respect to $\preceq$, $0 \in L$ is the minimal element of L with respect to $\preceq$, $0 \neq 1$, and for all a and b in L, $\sup \{a, b\}$ and $\inf \{a, b\}$ exist. For each a and b in $\mathfrak{L}$, let*

$a \sqcup b = \sup \{a, b\}$ *with respect to $\preceq$ and* $a \sqcap b = \inf \{a, b\}$ *with respect to $\preceq$.*

Then $\mathfrak{L} = \langle L, \sqcup, \sqcap, 1, 0 \rangle$ is a lattice.

Proof. Arguments for the axioms of zero element, meet associativity, meet commutativity, and meet absorption for a lattice will be given. The remaining axioms for a lattice—unit element, join associativity, join commutativity, and join absorption—follow by analogous arguments.

Zero element. It is immediate from the definitions of "minimal element", "sup", "inf", and Theorem 2.6 that the zero element axiom holds.

Meet commutativity. For all a and b in L, $a \sqcap b = \inf\{a, b\} = \inf\{b, a\} = b \sqcap a$.

Meet associativity. Let a, b, and c be arbitrary elements of L, and let u and v be defined as follows:

$$u = \inf\left\{a, \inf\left\{b, c\right\}\right\} \text{ and } v = \inf\left\{\inf\left\{a, b\right\}, c\right\}. \tag{2.1}$$

From the definition of "inf" and Equation 2.1 it follows that

$$u \preceq a \text{ and } u \preceq \inf\left\{b, c\right\},$$

and thus

$$u \preceq a, \ u \preceq b, \text{ and } u \preceq c. \tag{2.2}$$

From Equation 2.2 and the definition inf, it follows that

$$u \preceq \inf\left\{a, b\right\} \text{ and } u \preceq c,$$

and thus

$$u \preceq \inf\left\{\inf\left\{a, b\right\}, c\right\} = v. \tag{2.3}$$

From the definition of "inf" and Equation 2.1 it follows that

$$v \preceq \inf\left\{a, b\right\} \text{ and } v \preceq c,$$

and thus

$$v \preceq a, \ v \preceq b, \text{ and } v \preceq c. \tag{2.4}$$

From Equation 2.4 and the definition of inf, it follows that

$$v \preceq a \text{ and } v \preceq \inf\left\{b, c\right\}$$

and thus

$$v \preceq \inf\left\{a, \inf\left\{b, c\right\}\right\} = u. \tag{2.5}$$

Thus, because $\preceq$ is antisymmetric, it follows from Equations 2.3 and 2.5 that $u = v$. Because

$$a \sqcap (b \sqcap c) = u = v = (a \sqcap b) \sqcap c,$$

the axiom of meet associativity holds for $\mathfrak{L}$.

Meet absorption. Let a and b be arbitrary elements of L. Because $a \preceq \sup\{a, b\}$,

$$a \sqcap (a \sqcup b) = \inf\left\{a, \sup\left\{a, b\right\}\right\} = a,$$

and thus the axiom of meet absorption holds for $\mathfrak{L}$. $\square$

2.4 Distributive and Modular Lattices

Distributive and modular lattices are two of the most important classes of lattices. They have the right algebraic properties to be domains of probability functions. This section provides their definitions and a few of their basic properties. Chapters 4 and 5 provide some additional algebraic, topological, and probabilistic properties of distributive lattices.

Definition 2.11 (distributivity, $\sqcup$-distributivity). Let

$$\mathfrak{L} = \langle L, \sqcup, \sqcap, 1, 0 \rangle$$

be a lattice. Then $\mathfrak{L}$ is said to be *distributive* if and only if for all a, b, and c in L,

$$a \sqcap (b \sqcup c) = (a \sqcap b) \sqcup (a \sqcap c).$$

The dual of distributivity, which is called $\sqcup$-*distributivity*, is the property of satisfying the following identity for all a, b, c in L:

$$a \sqcup (b \sqcap c) = (a \sqcup b) \sqcap (a \sqcup c). \quad \square$$

Theorem 2.8. *Suppose* $\mathfrak{L} = \langle L, \sqcup, \sqcap, 1, 0 \rangle$ *is a lattice. Then the following two statements are equivalent.*

(1) $\mathfrak{L}$ *is distributive.*
(2) $\mathfrak{L}$ *is* $\sqcup$-*distributive.*

Proof. (1). Assume $\mathfrak{L}$ is distributive. Let a, b, and c be arbitrary elements of L. Then

$$(a \sqcup b) \sqcap (a \sqcup c) = [(a \sqcup b) \sqcap a] \sqcup [(a \sqcup b) \sqcap c] = a \sqcup (a \sqcap c) \sqcup (b \sqcap c)$$
$$= [a \sqcup (a \sqcap c)] \sqcup (b \sqcap c) = (a \sqcup (b \sqcap c)),$$

and thus Statement (2) holds.

Because Statement (2) is the dual of Statement (1), Statement (1) follows from the Duality Principle applied to the proof of Statement (1). $\quad \square$

Definition 2.12 (distributive inequality). Let $\mathfrak{L} = \langle L, \leq, \sqcup, \sqcap, 1, 0 \rangle$ be a lattice. The *distributive inequality* is said to hold for $\mathfrak{L}$ if and only if for all a, b, and c in L,

$$(a \sqcap b) \sqcup (a \sqcap c) \leq a \sqcap (b \sqcup c).$$

The dual of the distributive inequality, for all a, b, c in L,

$$(a \sqcup b) \sqcap (a \sqcup c) \geq a \sqcup (b \sqcap c),$$

is called the $\sqcup$-*distributive inequality*. $\quad \square$

The following theorem shows that the distributive and $\sqcup$-distributive inequalities are significant weakenings of, respectively, distributivity and $\sqcup$-distributivity.

Theorem 2.9. *Suppose $\mathfrak{L} = \langle L, \leq, \sqcup, \sqcap, 1, 0 \rangle$ is a lattice. Then the following two statements hold.*

(1) $\mathfrak{L}$ satisfies the distributive inequality.
(2) $\mathfrak{L}$ satisfies the $\sqcup$-distributive inequality.

   Proof. By Theorem 2.6, $x \sqcap y = \inf\{x, y\}$ and $x \sqcup y = \sup\{x, y\}$ for each x and y in L. Thus from

$$a \sqcap b \leq a \ \text{ and } \ a \sqcap c \leq a,$$

it follows that

$$(a \sqcap b) \sqcup (a \sqcap c) \leq a. \tag{2.6}$$

From

$$a \sqcap b \leq b \ \text{ and } \ a \sqcap c \leq c$$

it follows that

$$(a \sqcap b) \sqcup (a \sqcap c) \leq b \sqcup c. \tag{2.7}$$

And similarly, by Equations 2.6 and 2.7,

$$(a \sqcap b) \sqcup (a \sqcap c) \leq a \sqcap (b \sqcup c).$$

   Because Statement (2) is the dual of Statement (1), Statement (1) follows from the Duality Principle applied to the proof of Statement (1).   □

Definition 2.13 (modular, $\sqcup$-modular). Let $\mathfrak{L} = \langle L, \leq, \sqcup, \sqcap, 1, 0 \rangle$ be a lattice. Then $\mathfrak{L}$ is said to be *modular* if and only if for all a, b, c in L,

$$\text{if } a \leq b, \text{ then } (b \sqcap a) \sqcup (b \sqcap c) = b \sqcap (a \sqcup c),$$

$\mathfrak{L}$ is said to be $\sqcup$-*modular* if and only if for all a, b, c in L,

$$\text{if } b \leq a, \text{ then } (b \sqcup a) \sqcap (b \sqcup c) = b \sqcup (a \sqcap c).$$

Theorem 2.10. *Let $\mathfrak{L} = \langle L, \leq, \sqcup, \sqcap, 1, 0 \rangle$ be a lattice. Then $\mathfrak{L}$ is modular if and only if it is $\sqcup$-modular.*
   Proof. Theorem follows by duality.   □

Modularity is a generalization of distributivity. Von Neumann in the 1930s used complemented modular lattices as event spaces in his lattice investigations into foundations of quantum mechanics. Today, orthomodular lattices (Definition 1.18) are used for this instead of complemented modular lattices in algebraic treatments of quantum mechanics.

Lemma 2.2 (modular inequality). *Suppose* $\mathfrak{L} = \langle L, \leq, \sqcup, \sqcap, 1, 0 \rangle$ *is a lattice. Then the following three statements hold:*

(1) Modular inequality: *For all a, b, and c in L,*

$$\text{if } a \leq b \text{ then } a \sqcup (b \sqcap c) \leq (a \sqcup c) \sqcap b.$$

(2) $\mathfrak{L}$ *is modular if and only if for all a, b, and c in L,*

$$\text{if } a \leq b \text{ then } a \sqcup (b \sqcap c) = b \sqcap (a \sqcup c).$$

(3) $\mathfrak{L}$ *is not modular if and only if for some a, b, and c in L,*

$$\text{if } a \leq b \text{ then } a \sqcup (b \sqcap c) < b \sqcap (a \sqcup c).$$

Proof. Statement (1). Assume $a \leq b$. Because also $a \leq a \sqcup c$,

$$a \leq (a \sqcup c) \sqcap b. \tag{2.8}$$

Because $b \sqcap c \leq c \leq a \sqcup c$ and $b \sqcap c \leq b$, it follows that

$$b \sqcap c \leq (a \sqcup c) \sqcap b. \tag{2.9}$$

Then $a \sqcup (b \sqcap c) \leq (a \cup b) \cap c$ follow from Equations 2.8 and 2.9 and Theorem 2.6. Statement (2) immediately follows from Definition 2.13. And Statement (3) immediately follows from Statements (1) and (2). $\quad \square$

As shown in Theorem 2.11 below, the lattice described in Figure 2.1 provides a useful means for determining the modularity/non-modularity of a lattice.

Definition 2.14 (sublattice). Let $\mathfrak{L} = \langle L, \leq, \sqcup, \sqcap, 1, 0 \rangle$ be a lattice. Then a *sublattice* of $\mathfrak{L}$ is a lattice of the form,

$$\langle S, \leq, \sqcup, \sqcap, a, b \rangle,$$

where $S \subseteq L$ and $a \in S$ and $b \in S$. $\quad \square$

The following characterization of modular lattices is due to Dedekind (1900).

Theorem 2.11. *The following two statements are equivalent:*

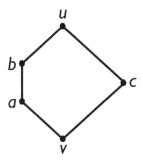

Fig. 2.1   The lattice N_5. In this diagram, $v < a < b < u$; $v < c < u$; $a \sqcap c = b \sqcap c = v$; $a \sqcup c = b \sqcup c = u$, etc.

(1) $\mathfrak{L}$ *is modular.*

(2) $\mathfrak{L}$ *does not have a* N_5 *sublattice, that is, does not have a sublattice that is isomorphic to the* N_5 *lattice shown in Figure 2.1.*

Proof.   Suppose Statement (1). Then it is immediate from Figure 2.1 that Statement (2) holds, that is, Figure 2.1 cannot hold, because

$$a \neq b,\ b = b \sqcap (a \sqcup c),\ \text{and}\ (b \sqcap a) \sqcup (b \sqcap c) = a\,.$$

Thus Statement (2) has been shown.

Suppose Statement (2). Suppose $\mathfrak{L}$ is not modular, that is, by Statement (3) of Lemma 2.2, let a, b, and c be elements of L such that

$$a \leq b\ \text{and}\ a \sqcup (b \sqcap c) < b \sqcap (a \sqcup c)\,. \tag{2.10}$$

A contradiction will be shown. This will be done by showing the five elements,

$$b \sqcap c,\ a \sqcup (b \sqcap c),\ b \sqcap (a \sqcup c),\ c,\ \text{and}\ a \sqcup c,$$

form the lattice N_5 of Figure 2.1

If $a \leq c$ or $c \leq b$, then Equation 2.10 cannot hold. Thus $a \nleq c$ and $c \nleq b$. Therefore, using Equation 2.10, we obtain

$$b \sqcap c < c < a \sqcup c\ \text{and}\ b \sqcap c < a \sqcup (b \sqcap c) < b \sqcap (a \sqcup c) < a \sqcup c\,. \tag{2.11}$$

Then it follows from Equation 2.10 that

$$a \sqcup c = (a \sqcup c) \sqcup c \leq [a \sqcup (b \sqcap c)] \sqcup c \leq [b \sqcap (a \sqcup c)] \sqcup c \leq (a \sqcup c) \sqcup c = a \sqcup c\,,$$

and thus,

$$a \sqcup c = [a \sqcup (b \sqcap c)] \sqcup c = [b \sqcap (a \sqcup c)] \sqcup c\,. \tag{2.12}$$

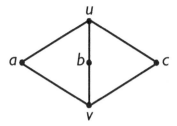

Fig. 2.2 The lattice M_3. In this diagram, $v < a < u$; $v < b < u$; $v < c < u$; $a \sqcap b = a \sqcap c = b \sqcap c = v$; $a \sqcup b = a \sqcup c = b \sqcup c = u$, etc.

Similarly, it follows from Equation 2.10 that

$$b \sqcap c = (b \sqcap c) \sqcap c \leq [a \sqcup (b \sqcap c)] \sqcap c \leq b \sqcap [(a \sqcup c) \sqcap c] \leq b \sqcap c = b \sqcap c,$$

and thus,

$$b \sqcap c = [a \sqcup (b \sqcap c)] \sqcap c = [b \sqcap (a \sqcup c)] \sqcap c. \tag{2.13}$$

Equations 2.11, 2.12, and 2.13 show that the five elements,

$$b \sqcap c, \ a \sqcup (b \sqcap c), \ b \sqcap (a \sqcup c), \ c, \ \text{and} \ a \sqcup c,$$

form the lattice N_5 of Figure 2.1. $\square$

The lattice described in Figure 2.2 similarly provides a useful means for determining the distributivity/non-distributivity of a modular lattice.

Birkhoff (1934) proved the following theorem.

Theorem 2.12. *Suppose $\mathfrak{L}$ is a modular lattice. Then the following two statements are equivalent:*

(1) $\mathfrak{L}$ is distributive.
(2) $\mathfrak{L}$ has no M_3 sublattice (Figure 2.2).

Proof. Suppose Statement (1). Then Statement (2) immediately follows.

Suppose Statement (2). Let x, y, and z be arbitrary elements of $\mathcal{L}$, and let

(i) $a = [x \sqcap (y \sqcup z)] \sqcup (y \sqcap z)$,
(ii) $b = [y \sqcap (z \sqcup x)] \sqcup (z \sqcap x)$,
(iii) $c = [z \sqcap (x \sqcup y)] \sqcup (x \sqcap y)$,
(iv) $v = (x \sqcap y) \sqcup (y \sqcap z) \sqcup (z \sqcap x)$,

(v) and $u = (x \sqcup y) \sqcap (y \sqcup z) \sqcap (z \sqcup x)$.

The basic idea of the proof is to show in succession that

- $a \sqcup b = a \sqcup c = b \sqcup c = v$,
- $a \sqcap b = a \sqcap c = b \sqcap c = u$,
- and $x \sqcap (y \sqcup z) = (x \sqcap y) \sqcup (x \sqcap z)$.

Applying the $\sqcup$-modular law to $[x \sqcap (y \sqcup z)] \sqcup [y \sqcap (z \sqcup x)]$ with $y \sqcap (z \sqcup x) \le y \sqcup z$ yields,

$$[x \sqcap (y \sqcup z)] \sqcup [y \sqcap (z \sqcup x)] = [(y \sqcup z) \sqcap [y \sqcap (z \sqcup x)] \sqcup x]. \qquad (2.14)$$

Because $x \le z \sqcup x$, it follows from Statement (2) of Lemma 2.2 that

$$[y \sqcap (z \sqcup x)] \sqcup x = (y \sqcup x) \sqcap [(z \sqcup x) \sqcup x] = (y \sqcup x) \sqcap (z \sqcup x). \qquad (2.15)$$

Thus, by substituting Equation 2.15 into the right of Equation 2.14 and the definition of v yields,

$$[x \sqcap (y \sqcup z)] \sqcup [y \sqcap (z \sqcup x)] = (y \sqcup z) \sqcap (y \sqcup x) \sqcap (z \sqcup x) = u. \qquad (2.16)$$

Because

$$y \sqcap z \le y \sqcap (z \sqcup x) \le b \text{ and } z \sqcap x \le x \sqcap (y \sqcup z) \le a,$$

it follows from Equation 2.16 and the definitions of a and b that

$$a \sqcup b = u.$$

Repetition of the above argument with the substitutions

$$x \mapsto y, \ y \mapsto z, \ z \mapsto z$$

yields,

$$[y \sqcap (z \sqcup x)] \sqcup [z \sqcap (x \sqcup y)] = u \text{ and } b \sqcup c = v,$$

and repetition with the substitutions

$$x \mapsto z, \ y \mapsto x, \ z \mapsto y$$

yields,

$$[z \sqcap (x \sqcup y)] \sqcup [x \sqcap (y \sqcup z)] = u \text{ and } c \sqcup a = v.$$

Thus it has been shown that

$$v = a \sqcup b = b \sqcup c = c \sqcup a. \qquad (2.17)$$

Because $y \sqcap z \le y \sqcup z$, application of Statement (2) of Lemma 2.2 to $a = [x \sqcap (y \sqcup z)] \sqcup (y \sqcap z)$ yields

$(i)'$ $a = [x \sqcup (y \sqcap z)] \sqcap (y \sqcup z)$.

In a similar manner the following are obtained:

$(ii)'$ $b = [y \sqcup (z \sqcap x)] \sqcap (z \sqcup x)$,
$(iii)'$ $c = [z \sqcup (x \sqcap y)] \sqcap (x \sqcup y)$.

Note that $(i)'$ is the dual of (i); $(ii)'$ is the dual of (ii); $(iii)'$ is the dual of (iii); and very importantly, v is the dual of u. Thus, because x, y, and z were chosen to be arbitrary elements of L, it follows from the duality principle and Equation 2.17 that

$$a \sqcap b = b \sqcap c = c \sqcap a = u. \tag{2.18}$$

Equations 2.17 and 2.18 show that if a, b, and c are distinct, then $\{u, a, b, c, v\}$ form a M_3 lattice. This is contrary to assumption. So suppose at least two of the a, b, c are identical. Without loss of generality, suppose $a = b$. Then by Equations 2.17 and 2.18 it follows that $u = v = a$ and thus, from the same equations, $c \sqcup a = c \sqcap a$. The latter can only happen if $a = c$. In other words, if a, b, and c are not distinct, then

$$a = b = c = u = v.$$

But then, by the definitions of a and v,

$$x \sqcap (y \sqcup z) \leq a = v \leq (x \sqcap y) \sqcup (x \sqcap z),$$

which together with the distributive inequality (Statement (1) of Theorem 2.9; Definition 2.12),

$$(x \sqcap y) \sqcup (x \sqcap z) \leq x \sqcap (y \sqcup z),$$

yields

$$x \sqcap (y \sqcup z) = (x \sqcap y) \sqcup (x \sqcap z). \quad \square$$

2.5 Lattices with Negative Operators

Definition 2.15 (negative operator). Let $\mathfrak{L} = \langle L, \sqcup, \sqcap, 1, 0 \rangle$ be a lattice. The $\vdash$ is said to be a *negative operator* on $\mathfrak{L}$ if and only if $\vdash$ is a function from L onto L such that $\vdash 1 = 0$ and $\vdash 0 = 1$. $\quad \square$

A special kind of negative operator is a complementation operation.

Definition 2.16 (complement). Let $\mathfrak{L} = \langle L, \sqcup, \sqcap, 1, 0 \rangle$ be a lattice and a be an element of L. Then b in L is said to be a *complement* of a, if and only if

$$b \sqcap a = 0 \quad \text{and} \quad b \sqcup a = 1 \,.$$

$\mathfrak{L}$ is said to be *complemented* if and only if each of its elements is complemented.   □

The symbol $\vdash$ is often used to denote a negative operator. When a lattice is complemented, it or the symbols $-$ and $\perp$ will also denote a complementation operator. Set-theoretic complementation will usually be denoted by $-$. In a few cases $-$ will denote a lattice complementation operator that is not set-theoretic complementation. Note that a may have many complements. In such situations $\vdash a$ or $-$ will denote only one of these. There are many situations in this book where $\vdash$ is a negative operator that is not a complementation operator. This is especially the case in situations where there are more than one complementation operator, or in situations where $\vdash$ is employed for some other purpose.

Definition 2.17 (boolean lattice). A *boolean lattice* is a complemented distributive lattice.   □

Note that in the literature boolean lattices, as defined here, are also called "boolean algebras". Throughout this book, the term "boolean algebra of events" is used to denote a special kind of boolean lattice with domain consisting of a collection of subsets of a nonempty set, its join and meet operations being, respectively, set-theoretic union and intersection, and its complementation operation being set theoretic complementation.

By convention, expressions like "$\mathfrak{L} = \langle L, \sqcup, \sqcap, \vdash, 1, 0 \rangle$ is a complemented lattice stands" for $\mathfrak{L} = \langle L, \sqcup, \sqcap, 1, 0 \rangle$ is a complemented lattice and $\vdash$ is a complementation operation on $\mathfrak{L}$. Similar conventions apply when $\vdash$ is some other specific kind of negative operator or just a negative operator.

Definition 2.18 (uniquely complemented lattices). A lattice $\mathfrak{L} = \langle L, \sqcup, \sqcap, \vdash, 1, 0 \rangle$ is said to be *uniquely complemented* if and only if it is complemented and for all a and b in L, if $a \sqcap b = 0$ and $a \sqcup b = 1$, then $b = \neg a$.   □

Not all complemented lattices are uniquely complemented. However, all boolean lattices are uniquely complemented.

Theorem 2.13. *Let* $\mathfrak{L} = \langle L, \sqcup, \sqcap, \ulcorner, 1, 0 \rangle$ *be a boolean lattice. Then* $\mathfrak{L}$ *is uniquely complemented.*

Proof. Suppose a is an arbitrary element of L and b is an element of L such that $a \sqcup b = 1$ and $a \sqcap b = 0$. It needs only to be shown that $\ulcorner a = b$. Because $\mathfrak{L}$ is distributive and $a \sqcap b = 0$,

$$b = b \sqcap (a \sqcup \ulcorner a) = (b \sqcap a) \sqcup (b \sqcap \ulcorner a) = (b \sqcap \ulcorner a),$$

and thus $b \leq \ulcorner a$. Similarly, because $\mathfrak{L}$ is distributive and $a \sqcup b = 1$, it follows from Theorem 2.8 that $\mathfrak{L}$ is $\sqcup$-distributive, and thus,

$$b = b \sqcup (a \sqcap \ulcorner a) = (b \sqcup a) \sqcap (b \sqcup \ulcorner a) = b \sqcup \ulcorner a,$$

and therefore,

$$\ulcorner a \sqcap b = \ulcorner a \sqcap (\ulcorner a \sqcup b) = \ulcorner a,$$

and thus $\ulcorner a \leq b$. This shows $\ulcorner a = b$. $\square$

Definition 2.19 (De Morgan Laws). Let $\mathfrak{L} = \langle L, \leq, \sqcup, \sqcap, \ulcorner, 1, 0 \rangle$ be a lattice. Then $\mathfrak{L}$ is said to satisfy *De Morgan's Laws* if and only if for all a and b in L,

- $\ulcorner (a \sqcap b) = (\ulcorner a) \sqcup (\ulcorner b)$, and
- $\ulcorner (a \sqcup b) = (\ulcorner a) \sqcap (\ulcorner b)$. $\square$

Theorem 2.14. *Let* $\mathfrak{L} = \langle L, \leq, \sqcup, \sqcap, \ulcorner, 1, 0 \rangle$ *be a complemented lattice. Then the following two statements are equivalent:*

(1) $\mathfrak{L}$ satisfies DeMorgan's laws.
(2) For all a and b in L,

(i) $a \leq b$ iff $\ulcorner b \leq \ulcorner a$, and
(ii) $\ulcorner\ulcorner a = a$.

Proof. Let a and b be arbitrary elements of L.

Assume Statement (1). To show $(2i)$, suppose $a \leq b$. Then $a = a \sqcap b$, and thus $\ulcorner a = \ulcorner (a \sqcap b) = \ulcorner a \sqcup \ulcorner b$. This implies that $\ulcorner b \leq \ulcorner a$, showing $2(i)$. To show $2(ii)$, note that from $\ulcorner a = \ulcorner a \sqcup \ulcorner a$, it follows from Statement (1) that

$$\ulcorner\ulcorner a = \ulcorner (\ulcorner a \sqcup \ulcorner a) = \ulcorner\ulcorner a \sqcap \ulcorner\ulcorner a = a \sqcap a = a,$$

showing $(2(ii))$.

Assume Statement (2). Because $a \leq a \sqcup b$, it follows that

$$\ulcorner (a \sqcup b) \leq \ulcorner a.$$

Similarly,

$$\ulcorner(a \sqcup b) \le \ulcorner b.$$

Thus,

$$\ulcorner(a \sqcup \ulcorner b) \le \ulcorner a \sqcap \ulcorner b. \tag{2.19}$$

Because $\ulcorner a \sqcap \ulcorner b \le \ulcorner a$ and $\ulcorner a \sqcap \ulcorner b \le \ulcorner b$, it follows from $(2(i))$ that

$$\ulcorner\ulcorner a \le \ulcorner(\ulcorner a \sqcap \ulcorner b) \text{ and } \ulcorner\ulcorner b \le \ulcorner(\ulcorner a \sqcap \ulcorner b),$$

which, by $\ulcorner\ulcorner a = a$ and $\ulcorner\ulcorner b = b$ yields,

$$a \sqcup b \le \ulcorner(\ulcorner a \sqcap \ulcorner b)$$

which implies that

$$\ulcorner\ulcorner(\ulcorner a \sqcap \ulcorner b) = \ulcorner a \sqcap \ulcorner b \le \ulcorner(a \sqcup b).$$

This shows that

$$\ulcorner(a \sqcup b) = \ulcorner a \sqcap \ulcorner b. \tag{2.20}$$

Then by Equation 2.20,

$$\ulcorner a \sqcup \ulcorner b = \ulcorner\ulcorner(\ulcorner a \sqcup \ulcorner b) = \ulcorner(\ulcorner\ulcorner a \sqcap \ulcorner\ulcorner b) = \ulcorner(a \sqcap b). \tag{2.21}$$

Because $\ulcorner a = \ulcorner a \sqcup \ulcorner a$, it follows from $(2(i))$ that

$$\ulcorner\ulcorner a = \ulcorner(\ulcorner a \sqcup \ulcorner a) = \ulcorner\ulcorner a \sqcap \ulcorner\ulcorner a = a \sqcap a = a. \tag{2.22}$$

Equations 2.20, 2.21, and 2.22 show that DeMorgan Laws, Statement (1), holds. $\square$

Although having a trivial proof, the following theorem is of some interest.

Theorem 2.15. *Suppose the lattice* $\mathfrak{L} = \langle L, \le, \sqcup, \sqcap, \ulcorner, 1, 0 \rangle$ *is a boolean algebra. Then De Morgan's Laws (Definition 2.19) hold.*

Proof. Let a and b be arbitrary elements of L. First it will be shown that $\ulcorner(a \sqcap b) = (\ulcorner a) \sqcup (\ulcorner b)$. Then by the distributivity of $\mathfrak{L}$,

$$(a \sqcap b) \sqcup [(\ulcorner a) \sqcap (\ulcorner b)] = [(a \sqcap b) \sqcap (\ulcorner a)] \sqcup [(a \sqcap b) \sqcap (\ulcorner b)]$$
$$= [a \sqcap (\ulcorner a) \sqcap b] \sqcup [a \sqcap b \sqcap (\ulcorner b)] = 0 \sqcup 0 = 0,$$
$$(a \sqcup b) \sqcap [(\ulcorner a) \sqcup (\ulcorner b)] = [(a \sqcup b) \sqcup (\ulcorner a)] \sqcap (a \sqcup b) \sqcup (\ulcorner b)]$$
$$= [a \sqcup (\ulcorner a) \sqcup b] \sqcap [a \sqcup b \sqcup (\ulcorner b)] = 1 \sqcap 1 = 1.$$

This shows that $(a \sqcap b) \sqcup [(\ulcorner a) \sqcap (\ulcorner b)]$ and $[(a \sqcap b) \sqcap (\ulcorner a)] \sqcup (a \sqcap b) \sqcap (\ulcorner b)]$ are complements of one another, and therefore by Theorem 2.13, $\ulcorner(a \sqcap b) = (\ulcorner a) \sqcup (\ulcorner b)$.

Applying duality to the above argument yields $\ulcorner(a \sqcup b) = (\ulcorner a) \sqcap (\ulcorner b)$. Thus Statement (1) has been shown. $\square$

2.6   Representation Theorems

This section provides lattice representation theorems for general, distributive lattices, and boolean lattices. A "lattice representation theorem" is a result that shows a type of lattice is isomorphic to a more familiar lattice, and, therefore, has the same algebraic properties as the familiar lattice.

Definition 2.20 (isomorphic). ϕ is said to be an *isomorphism from lattice* $\langle L, \leq, \sqcup, \sqcap, 1, 0\rangle$ *onto lattice* $\langle L', \leq', \sqcup', \sqcap', 1', 0'\rangle$ if and only if ϕ is a one-to-one function from L onto L' such that for all x and y in L,

- $x \leq y$ iff $\phi(x) \leq' \phi(y)$,
- $\phi(x \sqcup y) = \phi(x) \sqcup' \phi(y)$,
- $\phi(x \sqcap y) = \phi(x) \sqcap' \phi(y)$, and
- $\phi(1) = 1'$ and $\phi(0) = 0'$. $\square$

Definition 2.21 (event lattice, distributive and boolean algebras of events). An *event lattice* is a lattice of the form

$$\mathfrak{X} = \langle \mathcal{X}, \subseteq, \mathbb{U}, \mathbb{\cap}, X, \varnothing \rangle,$$

where $\mathcal{X}$ is a set of subsets of the nonempty set X. In probabilistic investigations, elements of $\mathcal{X}$ are usually called *events*. The following are two important classes of event lattices:

- Event lattices of the form $\langle \mathcal{X}, \subseteq, \cup, \cap, X, \varnothing \rangle$, are called *distributive algebras of events*, and
- complemented event lattices of the form $\langle \mathcal{X}, \subseteq, \cup, \cap, -, X, \varnothing \rangle$, where $-$ is the operation of set-theoretic complementation, are called *boolean algebras of events*. $\square$

An important way of understanding a lattice $\mathfrak{L}$ is to find an event lattice that is isomorphic to it. Theorem 2.16 below shows that each lattice is isomorphic to an event lattice. In this book, such event lattices are important for the development of probability theory on lattices, because the set-theoretic relationship between X and $\mathcal{X}$ is interpreted as a relationship of occurrence, that is, elements of X are interpreted as states of the world, and for A in $\mathcal{X}$, A is interpreted as occurring if the real state of the world is in A and as non-occurring otherwise.

Theorem 2.16. *Let $\mathfrak{L} = \langle L, \leq, \sqcup, \sqcap, 1, 0\rangle$ be a lattice. Then there exists an isomorphism from $\mathfrak{L}$ onto an event lattice.*
    Proof. Let

- $J_0 = \varnothing$.
- $J_a = \{b \mid 0 < b \leq a\} \cup \{\varnothing\}$ for each $a \in L$ such that $a \neq 0$,
- $X = J_1 = L$, and
- $\mathcal{X} = \{J_a \mid a \in L\}$.

Let φ be the function from L onto X such that for each a in L, $\varphi(a) = J_a$. It will be shown that φ is an isomorphism. Let a and b be arbitrary elements of L.

It is immediate from the definitions of φ and $\mathcal{X}$ that φ is onto $\mathcal{X}$. Suppose $\varphi(a) = \varphi(b)$. If $a = 0$, then it immediately follows that $b = 0$ and vice versa. Thus suppose $a \neq 0$ and $b \neq 0$. Then $a \in \varphi(b)$ and $b \in \varphi(a)$. Then by the definitions of φ and J_x, $a \leq b$ and $b \leq a$, and thus $a = b$. This establishes that φ is a one-to-one function from $\mathfrak{L}$ onto $\mathfrak{X}$.

Part 1. Suppose $a \leq b$. If $a = 0$, then it is immediate that $\varnothing = \varphi(a) \subseteq \varphi(b)$. Thus suppose $a \neq 0$. Let c be an arbitrary element of $\varphi(a)$. Then $c \leq a$. Because $\leq$ is transitive and $a \leq b$, it follows that $c \leq b$. Therefore, by the definition of J_x, $c \in \varphi(b)$. Thus, because c is an arbitrary element of $\varphi(a)$, $\varphi(a) \subseteq \varphi(b)$.

Part 2. Suppose $\varphi(a) \subseteq \varphi(b)$. If $a = 0$, then $a \leq b$. So suppose $a \neq 0$. Then, because $a \in \varphi(a)$, it follows that $a \in \varphi(b)$. Thus, by the definition of J_x, $a \leq b$. Parts 1 and 2 show that

$$a \leq b \ \text{ iff } \ \varphi(a) \subseteq \varphi(b).$$

Part 3. If $a = 0$, then

$$\varphi(a \sqcap b) = \varnothing = \varnothing \cap \varphi(b) = \varphi(a) \cap \varphi(b),$$

and similarly for $b = 0$. Suppose $a \neq 0$ and $b \neq 0$. Let c be an arbitrary element of L such that $c \leq (a \sqcap b)$. (Such a c exists, because $0 \leq a \sqcap b$.) Thus $c \leq a$ and $c \leq b$. Therefore, $c \in \varphi(a)$ and $c \in \varphi(b)$, from which it follows that $c \in \varphi(a) \cap \varphi(b)$.

Part 4. Suppose w is an arbitrary element of $\varphi(a) \cap \varphi(b)$. Then $w \in \varphi(a)$ and $w \in \varphi(b)$. By the definitions of φ and J_x, it then follows that $w \leq a$ and $w \leq b$, and thus that $w \leq (a \sqcap b)$. Parts 3 and 4 show that

$$\varphi(a \sqcap b) = \varphi(a) \cap \varphi(b).$$

Because φ is a function from L onto $\mathcal{X}$, define the binary operation $\uplus$ on $\mathcal{X}$ as follows: For all e and f in L,

$$\varphi(e) \uplus \varphi(f) = \varphi(e \sqcup f).$$

It follows from the definitions of φ, J_0 and J_1 that $\varphi(0) = J_0 = \varnothing$ and $\varphi(L) = J_1 = X$. This together with the above establishes that φ is an isomorphism from $\mathfrak{L} = \langle L, \leq, \sqcup, \sqcap, 1, 0 \rangle$ onto $\mathfrak{X} = \langle \mathcal{X}, \subseteq, \uplus, \cap, X, \varnothing \rangle$. $\square$

The following theorem is due to Stone (1936). The proof given here is a simplified version of Stone's proof due to Frink (1941).

Theorem 2.17. *Let $\mathfrak{L} = \langle L, \sqcup, \sqcap, \multimapdotinv, 1, 0 \rangle$ be a boolean lattice. Then there exists an isomorphism from $\mathfrak{L}$ onto a boolean algebra of events.*

Proof. Note that for all a and b in L,

$$b \sqcap (\multimapdotinv a) = 0 \quad \text{iff} \quad \multimapdotinv \multimapdotinv a \leq b \quad \text{iff} \quad a \leq b \quad \text{iff} \quad a \sqcap b = a. \qquad (2.23)$$

By definition, a *point* is a set P of elements of L such that

(i) $0 \notin P$,
(ii) if $a \in P$ and $b \in P$, then $(a \sqcap b) \in P$, and
(iii) P is maximal with respect to properties (i) and (ii).

Note that (iii) implies that $1 \in P$.
For each a in L, let

$$R_a = \{P \,|\, P \text{ is a point and } a \in P\}.$$

R_a will be the event in the boolean algebra of events that corresponds to a. The following two lemmas are needed.

Lemma α: Let P be point, $a \in L$, and $b \in P$. Then if $(a \sqcap b) \in P$ then $a \in P$. Proof: Suppose $(a \sqcap b) \in P$ and $a \notin P$. A contradiction will be shown. Then

$$P_1 = P \cup \{p \sqcap a \,|\, p \in P\}$$

P_1 satisfies (ii), and, because $1 \in P$, $P \subset P_1$. P_1 also satisfies (i), because if $p \sqcap a$ were 0 for some p in P, then $0 = (p \sqcap a) \sqcap b = p \sqcap (a \sqcap b)$, which contradicts the assumption $(a \sqcap b) \in P$ and (ii). Therefore P is not a point, because it doe not satisfy (iii).

Lemma β: Let $a \in L$ and $a \neq 0$. Then for some point P, $a \in P$. Proof. Let S be a set of elements of L that has a as an element and satisfies (i) and (ii). Then $\subseteq$ is a partial ordering on S. Each $\subseteq$-totally ordered subsystem $\mathcal{T}$ of S has an upper bound in S, namely $\bigcup \mathcal{T}$. Thus by Zorn's lemma, S has a $\subseteq$-maximal element P.

Let $\mathcal{R} = \{R_a \,|\, a \in L\}$ and $\mathcal{P} = \{P \,|\, P \text{ is a point}\}$. To show that the correspondence between elements a of L and their representative sets R_a is an isomorphism, it is sufficient to show the following for all a and b in L:

(1) $R_{a \sqcap b} = R_a \cap R_b$;

(2) $R_{\vdash a} = -R_a$ (where $-$ is set-theoretic complementation with respect to $\mathcal{P}$); and

(3) if $R_a = R_b$, then $a = b$.

(The "sufficiency" follows, because by DeMorgan's Laws, $\sqcup$ is definable in terms of $\sqcap$ and $\vdash$ (and a similar reason for $\cup$) and $\leq$ is definable in terms of $\sqcap$ (and a similar reason for $\subseteq$).)

(1). If $P \in R_a$ and $P \in R_b$, then, by (ii), $P \in R_{a \sqcap b}$. Also, if $P \in R_{a \sqcap b}$, then, by Lemma α, $P \in R_a$ and $P \in R_b$.

(2). $R_a \cup R_{\vdash a} = \mathcal{P}$, because if an element P of $\mathcal{P}$ is not in $R_{\vdash a}$, then there is $b \in P$ such that $(\vdash a) \sqcap b = 0$, for otherwise $\vdash a$ and products of the form $(\vdash a) \sqcap p$, $p \in P$ could be added to P, contradicting the maximality of P. Thus by Equation 2.23, $a \sqcap b = b$, and therefore $(a \sqcap b) \in P$. It then follows from Lemma β that $a \in P$ and $P \in R_a$. To show that $R_a \sqcap R_{\vdash a} = \varnothing$, suppose $P \in (R_a \cap R_{\vdash a})$. A contradiction will be shown. Then $[a \sqcap (\vdash a)] \in P$, that is $0 \in P$, contradicting (i).

(3). Suppose $a \neq b$. It will be shown that $R_a \neq R_b$. From $a \neq b$ it follows that either $a \sqcap (\vdash b) \neq 0$ or $(\vdash a) \sqcap b \neq 0$, because otherwise, by Equation 2.23, $a = a \sqcap b = b$. If $a \sqcap (\vdash b) \neq 0$, then by Lemma β, $a \sqcap (\vdash b)$ is in some point P. By Lemma α, $a \in P$. But $b \notin P$, thus $R_a \neq R_b$. $\quad\square$

2.7 Miscellaneous Results

The proofs of certain theorems in a later chapter require the concept of "relative complement" and Lemmas 2.3 and 2.4 below.

Definition 2.22 (relative complement). Let $\mathfrak{L} = \langle \mathcal{L}, \leq, \sqcup, \sqcap, 1, 0 \rangle$ be a lattice and a and b be arbitrary elements of $\mathcal{L}$ such that $a \leq b$. Then by definition,

$$[a, b] = \{x \mid x \in \mathcal{L} \text{ and } a \leq x \leq b\}.$$

It is immediate that $[a, b]$ is a sublattice of $\mathfrak{L}$, and $[a, b]$ is said to be *relatively complemented* if and only if it is complemented as a lattice. For each x in $[a, b]$, an element y in $[a, b]$ such that

$$a = x \sqcap y \text{ and } b = x \sqcup y$$

is called the *relative complement of x with respect to $[a, b]$*.

Lemma 2.3. Let $\mathfrak{L} = \langle \mathcal{L}, \sqcup, \sqcap, 1, 0 \rangle$ be a complemented modular lattice and a, b, x, and t be elements of $\mathcal{L}$ such that $a \leq x \leq b$ and t is a complement *of* x. *Then*

$$y = (a \ \sqcup t) \sqcap b$$

is the relative complement of x *in the lattice* $[a, b]$.

Proof. Because $a \leq x$ and $t \sqcap x = 0$, it follows by the modularity of $\mathfrak{L}$ that

$$x \sqcap y = (a \ \sqcup t) \sqcap b \sqcap x = (a \ \sqcup t) \sqcap x \qquad (2.24)$$
$$= (a \sqcap x) \sqcup (t \sqcap x) = a \sqcup (t \ \sqcap \ x) = a \sqcup 0 = a \,.$$

Because $a \leq x \leq b$ and $x \sqcup t = 1$, it then follows from $x \sqcup a = x$ (Theorem 2.6) and the modularity of $\mathfrak{L}$ that

$$x \sqcup y = x \sqcup [(a \ \sqcup t) \sqcap b] = x \sqcup [(a \sqcap b) \sqcup (t \sqcap b)] \qquad (2.25)$$
$$= x \sqcup [a \sqcup (t \sqcap b)] = (x \sqcup a) \sqcup (t \sqcap b) = x \sqcup (t \sqcap b)$$
$$= (x \sqcap b) \sqcup (t \sqcap b) = (x \sqcup t) \sqcap b = 1 \sqcap b = b \,.$$

It follows from Equations 2.24 and 2.25 that y is the relative complement of x in the lattice $[a, b]$.

The following is the converse of Lemma 2.3. The proof follows Szász (1963).

Lemma 2.4. *Let* $\mathfrak{L} = \langle \mathcal{L}, \sqcup, \sqcap, 1, 0 \rangle$ *be a complemented modular lattice and* a, b, *and* x *be elements of* $\mathcal{L}$ *such that* $a \leq x \leq b$. *Then for each relative complement* y *of* x *in* $[a, b]$ *there exists a complement* t *of* x *in* $\mathfrak{L}$ *such that*

$$y = (a \sqcup t) \sqcap b \,.$$

Proof (Szász, 1963, p. 112). Let y be an arbitrary relative complement of x in $[a, b]$. Let

- v be a relative complement of a in $[0, y]$,
- u be a relative complement of b in $[y, 1]$,
- and t the relative complement of y in $[v, u]$.

Then by the definition of "relative complement" the following equations hold:

$$v \leq t \leq u \qquad (2.26)$$
$$x \sqcap y = a \quad \text{and} \quad x \sqcup y = b \qquad (2.27)$$
$$a \sqcap v = 0 \quad \text{and} \quad a \sqcup v = y \qquad (2.28)$$
$$b \sqcap u = y \quad \text{and} \quad b \sqcup u = 1 \qquad (2.29)$$
$$y \sqcap t = v \quad \text{and} \quad y \sqcup t = u \,. \qquad (2.30)$$

The following argument shows that t is a complement of x in $\mathfrak{L}$: By assumption, $x = x \sqcap b$. By Equations 2.26, 2.29, 2.27, 2.30, and 2.28,

$$x \sqcap t = (x \sqcap b) \sqcap (u \sqcap t) = x \sqcap (b \sqcap u) \sqcap t$$
$$= x \sqcap y \sqcap t = (x \sqcap y) \sqcap (y \sqcap t) = a \sqcap v = 0.$$

And similarly, using $x \sqcup a = a$ (Theorem 2.6),

$$x \sqcup t = (x \sqcup a) \sqcup (v \sqcup t) = x \sqcup (a \sqcup v) \sqcup t$$
$$= x \sqcup y \sqcup t = (x \sqcup y) \sqcup (y \sqcup t) = b \sqcup u = 1.$$

t also satisfies,

$$y = (a \sqcup t) \sqcap b,$$

because by Equations 2.26, 2.28, 2.30, and 2.29,

$$(a \sqcup t) \sqcap b = (a \sqcup (v \sqcup t)) \sqcap b = [(a \sqcup v) \sqcup t] \sqcap b$$
$$= (y \sqcup t) \sqcap b = u \sqcap b = y.$$

2.8   Unique Complementation and Distributivity

Boolean lattices are more than distributive and complemented: they are distributive and *uniquely* complemented:

Theorem 2.18. *Each boolean lattice is uniquely complemented.*

   Proof. Let $\mathfrak{L} = \langle L, \leq, \sqcup, \sqcap, 1, 0 \rangle$ be a distributive lattice and a be an arbitrary element of $\mathcal{L}$. Suppose b and c are complements of a. Then

$$b = b \sqcap 1 = b \sqcap (a \sqcup c) = (b \sqcap a) \sqcup (b \sqcap c) = 0 \sqcup (c \sqcap b) = b \sqcap c.$$

This shows $b \leq c$. Similarly,

$$c = c \sqcap (a \sqcup b) = (c \sqcap a) \sqcup (c \sqcap b) = c \sqcap b,$$

showing $c \leq b$. Thus, because $b \leq c$ and $c \leq b$, it follows that $b = c$.   □

In 1885, the Philosopher and logician Peirce communicated to other researchers a result of his showing that uniquely complemented lattices were boolean. His proof, however, had an error. But the idea that every uniquely complemented lattice was boolean would turn out to be a theorem of lattice theory persisted among lattice theorists in the early days of lattice theory until Dilworth (1945) produced a counterexample. However, like in Theorem 4.10 below, when various desirable conditions were added to unique complementation, it was shown that boolean lattices resulted. Combined, such theorems revealed a limited range of options available for event spaces founded on uniquely complemented lattices. This section shows two theorems that formulate boolean lattices in terms of concepts that emphasize the role of complementation. These theorems, combined with Theorem 4.10, provide considerable insight into the range of generalizations of classical probability theory.

The first theorem, Theorem 2.19, emphasizes two classical properties of complementation in boolean lattices—De Morgan's Laws and unique complementation. It shows that lattices satisfying these properties are boolean. Birkhoff and von Neumann (1936) developed a probability theory for the logical structure of quantum mechanics that kept De Morgan's Laws but gave up unique complementation. As a result of Theorem 2.19, their development of lattice of quantum propositions had to be non-distributive. The probability theory they developed for their lattice of propositions had a probability function that disobeyed a fundamental property of classical probability theory that required the probability of the join of disjoint events to be the sum of their probabilities. Because of this they appear to dismiss the idea that the complementation operation of a boolean algebra could be altered instead of distributivity to achieve a logic with a probability theory for science:

> The main difference seems to be that whereas logicians have usually assumed that the properties [given by the identity $--a = a$, the De Morgan Laws, and the consequence of De Morgan's Laws, if $a \leq b$, then $-b \leq -a$, stated in Theorem 2.15] of negation were the ones least able to withstand a critical analysis, the study of mechanics points to the *distributive identity* [i.e., distributivity, Definition 2.11] as the weakest link in the algebra of logic." *(Birkhoff & von Neumann, 1936, p. 837)*

However, as results of this book will demonstrate, their pessimism regarding the role of negation was misguided.

The second theorem is a modification of an erroneous "theorem" of Peirce from 1880, or more accurately, a modification of the argument he presented for his "theorem". In 1904, Peirce later asked the mathematician Huntington to include a restatement of his result using the same argument in an article on axiomatizations of boolean algebras that Huntington was preparing for publication (Huntington, 1904). Peirce's restatement included an additional and somewhat weird condition, Huntington included Peirce's request in his article along with some of Huntington's own suggestions for replacing Peirce's weird condition with more intuitive ones while retaining most of Peirce's proof. Theorem 2.21 below is one of Huntington's ideas with a modified proof. It states that every complemented and pseudo complemented lattice is distributive. Huntington conjectured that every uniquely complemented lattice was boolean. The conjecture subsequently received repeated partial confirmation by results showing that unique complementation and some other natural condition on a lattice produced a boolean lattice. This led to the expectation in the 1930s that Huntington's conjecture would likely be shown. However, Dilworth (1945) showed the conjecture to be false.

Lemma 2.5. *Suppose* $\mathfrak{L} = \langle L, \sqcup, \sqcap, -, 1, 0 \rangle$ *is a uniquely complemented lattice, De Morgan's Laws hold, and a and b are arbitrary elements of L such that $a \le b$. Then*

$$a = (a \sqcup -b) \sqcap b \quad and \quad b = a \sqcup (-a \sqcap b).$$

Proof. Because by hypothesis $a \le b$, it follows that

$$a \sqcup [(-a) \sqcap b] \le b.$$

Thus,

$$1 = -(a \sqcup [-a \sqcap b]) \sqcup (a \sqcup [-a \sqcap b]) \le -(a \sqcup [-a \sqcap b]) \sqcup b. \qquad (2.31)$$

By De Morgan's Laws and the fact that De Morgan's Laws implies double complementation (Theorem 2.15),

$$-[a \sqcup -(a \sqcap b)] \sqcap b = [(-a) \sqcap (-[-a \sqcap b])] \sqcap b \qquad (2.32)$$

$$= [-a \sqcap [a \sqcup -b]] \sqcap b$$

$$= [-a \sqcap b] \sqcap [a \sqcup -b]$$

$$= [-a \sqcap b] \sqcap [--a \sqcup -b]$$

$$= [-a \sqcap b] \sqcap -[-a \sqcap b] = 0.$$

From Equations 2.31 and 2.32 it follows that b and $a \sqcup -a \sqcap b$ are complements of $-[a \sqcup -a \sqcap b]$. Thus because, by hypothesis, $-$ is the unique complementation operator on $\mathfrak{L}$,

$$b = (a \sqcup -a \sqcap b).$$

Because, by hypothesis, De Morgan's Laws hold and $a \le b$, it follows from Theorem 2.15 that $-b \le -a$. Thus application of the above argument to $-b$ and $-a$ yields,

$$-a = -b \sqcup [--b \sqcap -a],$$

which by De Morgan's Laws and double complementation yield,

$$a = --a = -[-b \sqcup (--b \sqcap -a)] = b \sqcap -(--b \sqcap -a) = (a \sqcup -b) \sqcap b. \qquad \square$$

Theorem 2.19. *Suppose* $\mathfrak{L} = \langle L, \sqcup, \sqcap, -, 1, 0 \rangle$ *is a uniquely complemented lattice and De Morgan's Laws hold. Then* $\mathfrak{L}$ *is boolean.*

Proof. The theorem will be shown by showing that $\mathfrak{L}$ has no N_5 and

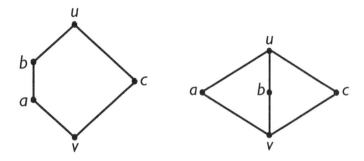

Fig. 2.3   Lattices N_5 and M_3. (Left N_5 and right M_3.)

no M_3 sublattices. Using the labeling of the elements of the N_5 and M_3 lattices in Figure 2.3 it follows that in both lattices,

$$a \neq b \tag{2.33}$$

and

$$a \sqcup c = u = b \sqcup c \quad \text{and} \quad a \sqcap b = v = b \sqcap c. \tag{2.34}$$

Thus to show theorem, it needs only to be shown that Equations 2.33 and 2.34 imply a contradiction. This will be done by showing that $-u \sqcup (c \sqcap -v)$ has both a and b as complements, and thus by the unique complementation of $\mathfrak{L}$, $a = b$, contradicting Equation 2.33.

Because $v \leq c$, it follows from Lemma 2.5 that

$$v \sqcup (c \sqcap -v) = c.$$

Note that because $v \leq a$ and $v \leq b$,

$$a = a \sqcup v \quad \text{and} \quad b = b \sqcup v.$$

Thus,

$$a \sqcup -u \sqcup (c \sqcap -v) = a \sqcup (c \sqcap -v) \sqcup -u = a \sqcup v \sqcup (c \sqcap -v) \sqcup -u \tag{2.35}$$

$$= a \sqcup [v \sqcup (c \sqcap -v)] \sqcup -u = a \sqcup c \sqcup -u$$

$$= (a \sqcup c) \sqcup -u = (a \sqcup c) \sqcup -(a \sqcup c) = 1$$

$$= (b \sqcup c) \sqcup -(b \sqcup c) = b \sqcup c \sqcup -u$$

$$= b \sqcup [v \sqcup (c \sqcap -v)] \sqcup -u = (b \sqcup v) \sqcup -u \sqcup (c \sqcap -v)$$

$$= b \sqcup -u \sqcup (c \sqcap -v).$$

Because $c \sqcap -v \leq u$, it follows from Lemma 2.5 that

$$u \sqcap [-u \sqcup (c \sqcap -v)] = c \sqcap -v.$$

Note that because $a \leq u$ and $b \leq u$,

$$a = a \sqcap u \quad \text{and} \quad b = b \sqcap u.$$

Thus,

$$\begin{aligned}
a \sqcap [-u \sqcup (c \sqcap -v)] &= a \sqcap u \sqcap [-u \sqcup (c \sqcap -v)] \qquad\qquad (2.36)\\
&= a \sqcap [u \sqcap (-u \sqcup (c \sqcap -v))] = a \sqcap [c \sqcap -v]\\
&= (a \sqcap c) \sqcap -v = 0 = (b \sqcap c) \sqcap -v\\
&= (b \sqcap u) \sqcap c \sqcap -v = b \sqcap [u \sqcap (c \sqcap -v)]\\
&= b \sqcap [-u \sqcup (c \sqcap -v)].
\end{aligned}$$

Equations 2.35 and 2.36 show that a and b are complements of $-u \sqcup (c \sqcap -v)$. Because $\mathfrak{L}$ is uniquely complemented, it follows that $a = b$. $\quad\square$

Definition 2.23 (pseudo complementation). Let $\mathfrak{L} = \langle \mathcal{L}, \leq, \sqcup, \sqcap, 1, 0 \rangle$ be a lattice and a, b, and c be arbitrary elements of $\mathcal{L}$. Then the following definitions hold:

- c is said to be the *pseudo complement of a* if and only if for each x in $\mathcal{L}$,

$$x \leq c \text{ iff } a \sqcap x = 0.$$

- $\mathfrak{L}$ is said to be *pseudo complemented* if and only if each element of $\mathcal{L}$ is pseudo complemented. $\quad\square$

Pseudo complemented lattices are discussed much more thoroughly in Chapter 3.

Theorem 2.20. *Let $\mathfrak{L} = \langle L, \sqcup, \sqcap, \vdash, 1, 0 \rangle$ be a pseudo complemented lattice and a and b be arbitrary elements of L. Then the following three statements hold.*

(1) If $b \leq a$ then $\vdash a \leq \vdash b$.
(2) $a \leq (\vdash\vdash a)$.
(3) $(\vdash a) = (\vdash\vdash\vdash a)$.

Proof. (1). Suppose $b \leq a$. Then, because $a \sqcap (\vdash a) = 0$, $b \sqcap (\vdash a) = 0$, and thus by the definition of "pseudo complement", $(\vdash a) \leq (\vdash b)$.

(2). By the definition of "pseudo complement", $(\vdash a) \sqcap a = 0$, and thus, again by the definition of "pseudo complement", $a \leq (\vdash\vdash a)$.

(3). By Statement (2), $a \le (\ulcorner\ulcorner a)$. Therefore by Statement (1),

$$(\ulcorner\ulcorner\ulcorner a) \le (\ulcorner a).$$

But by Statement (2),

$$\ulcorner a \le (\ulcorner\ulcorner) \ulcorner a = (\ulcorner\ulcorner\ulcorner a).$$

Thus, $(\ulcorner a) = (\ulcorner\ulcorner\ulcorner a)$.

Theorem 2.21 (Huntington's Theorem). *Suppose*

$$\mathfrak{L} = \langle \mathcal{L}, \sqcup, \sqcap, -, 1, 0 \rangle$$

is a lattice where $-$ *is both a complementation operation and a pseudo complementation operation. The* $\mathfrak{L}$ *is boolean.*

Proof. Let a, b, and c be arbitrary elements of L. The proof of theorem uses the following three lemmas.

Lemma 2.6. *The following two statements hold:*

(1) $--a = a$.
(2) $-a \sqcap -b \le -(a \sqcup b)$.

Proof. (1). Because a is a complement of $-a$ it is, by hypothesis, also a pseudo complement of a. Thus, because $--a \sqcap -a = 0$, $--a \le a$ by the definition of a being the pseudo complement of $-a$. By Statement (2) of Theorem 2.20, $a \le --a$. Thus $a = --a$.

(2). $-a \sqcap -b \le -a$. Then by Statement (1) of Theorem 2.20, $a \le -(-a \sqcap -b)$. Similarly, $b \le -(-a \sqcap -b)$. Therefore, $a \sqcup b \le -(-a \sqcap -b)$, which by Statement (1) of this Theorem and Statement (1) of Theorem 2.20 yields,

$$-a \sqcup -b = --(-a \sqcup -b) \le -(a \sqcup b). \quad \square$$

Lemma 2.7. *Suppose* $a \not\le b$. *Then there exists* x *such that*

$$x \ne 0, \ x \le a, \ and \ x \le -b.$$

Proof. It will first be shown by contradiction that $a \sqcap -b \ne 0$. Suppose $a \sqcap -b = 0$. Because $-$ is a pseudo complementation, $-b \le -a$, and thus by Statement (2) of Theorem 2.15, $--a \le --b$. Because $-$ is also a complementation operation, it then follows by Theorem 2.15 that $a \le b$, contrary to hypothesis.

Because $a \sqcap -b \ne 0$, there must be an element $x \ne 0$ such that $x \le a$ and $x \le -b$.

Lemma 2.8. $a \sqcap (b \sqcup c) \leq b \sqcup (a \sqcap c)$.

Proof. Suppose $a \sqcap (b \sqcup c) \not\leq b \sqcup (a \sqcap c)$. A contradiction will be shown.
It follows from Lemma 2.7 and the definition of "$\leq$" that $x \neq 0$ in L can
be found such that

$$x \leq a \sqcap (b \sqcup c) \tag{2.37}$$

and

$$x \leq -[b \sqcup (a \sqcap c)], \tag{2.38}$$

which by Lemma 2.6 yields

$$b \sqcup (a \sqcap c) \leq -x. \tag{2.39}$$

It follows from Equation 2.37 that

$$x \leq a \text{ and } x \leq (b \sqcup c). \tag{2.40}$$

It follows from Equation 2.39 and $-$ being a pseudo complement (and
Lemma 2.6) that

$$x \sqcap [b \sqcup (a \sqcap c)] = 0,$$

and thus that

$$x \sqcap b = 0 \text{ and } x \sqcap (a \sqcap c) = 0,$$

which, by $-$ being an operation of pseudo complementation, yields

$$b \leq -x \text{ and } a \sqcap c \leq -x. \tag{2.41}$$

From $a \sqcap c \leq -x$ (Equation 2.41) and $x \leq a$ (Equation 2.40), it follows that

$$x \not\leq c, \tag{2.42}$$

because, $x \leq c$ together with $x \leq a$ yields $x \leq a \sqcap c$ and thus $-(a \sqcap c) \leq -x$,
which is impossible by Equation 2.41. Therefore, by Lemma 2.7, let $y \neq 0$
in L be such that

$$y \leq x \text{ and } y \leq -c. \tag{2.43}$$

Then, by Statement (1) of Theorem 2.6,

$$c = --c \leq -y. \tag{2.44}$$

It follows from $y \leq x$ (Equation 2.43) and $x \leq -b$ (a consequence of
Equation 2.41) that

$$y \leq -b. \tag{2.45}$$

From $c \leq y$ (Equation 2.44) and $b \leq y$ (a consequence of Equation 2.44), it follows that

$$c \sqcup b \leq y. \tag{2.46}$$

It follows from $y \leq -b$ (Equation 2.45), $y \leq -c$ (a consequence of Equation 2.44), and Statement (2) of Theorem 2.6 that

$$y \leq (-b \sqcap -c) \leq -(b \sqcup c). \tag{2.47}$$

It follows from $y \leq x$ (Equation 2.43) and $x \leq b \sqcup c$ (Equation 2.40) that

$$y \leq b \sqcup c. \tag{2.48}$$

Equations 2.47 and 2.48 and the choice of $y \neq 0$ contradict $-$ being a complementation operation. $\square$

Proof of Theorem 2.21. By Lemma 2.8,

$$a \sqcap (b \sqcup c) \leq b \sqcup (a \sqcap c). \tag{2.49}$$

Therefore, by $\sqcap$ producing $\leq$-greatest lower bounds,

$$a \sqcap [a \sqcap (b \sqcup c)] \leq a \sqcap [b \sqcup (a \sqcap c)]. \tag{2.50}$$

By Lemma 2.8,

$$a \sqcap [b \sqcup (a \sqcap c)] = a \sqcap [(a \sqcap c) \sqcup b] \leq (a \sqcap c) \sqcup (a \sqcap b) = (a \sqcap b) \sqcup (a \sqcap c). \tag{2.51}$$

It follows from Equations 2.49, 2.50 and 2.51 that

$$a \sqcap (b \sqcup c) = a \sqcap [a \sqcap (b \sqcup c)] \leq (a \sqcap b) \sqcup (a \sqcap c).$$

However, by the distributive inequality (Theorem 2.9, which holds for all lattices),

$$(a \sqcap b) \sqcup (a \sqcap c) \leq a \sqcap (b \sqcup c),$$

and thus $\mathfrak{L}$ is distributive. $\square$

Theorem 4.9 of Chapter 4 says that each L-probabilistic uniquely complemented lattice is boolean. Thus, to generalize boolean lattices for possible applications involving L-probability, one must look beyond uniquely complemented lattices. Theorem 2.21 suggests two approaches for such generalization: (i) Investigate lattices that are pseudo complemented but not complemented, and (ii) investigate lattices that are complemented but not a pseudo complemented. It follows by Theorem 4.11 of Chapter 4 that for lattices of either approach to be L-probabilistic, they must be modular. For (ii) to hold, such modular lattices must be non-distributive (because, if distributive, they would be boolean and thus have their complementation operators also be pseudo complementation operators, contrary to (ii)). Non-distributive, complemented modular lattices have been much studied in mathematics because of their connection to projective geometry. In the literature, real valued L-probability functions on them usually have been investigated as a measure of dimensionality rather than as a measure of uncertainty. The existence of M_3 sublattices for them (Theorem 2.12, Figure 2.2) appear to cause extreme difficulties in being able to interpret L-probability functions on them as measures of uncertainty. Instead, to measure uncertainty for such lattices in a way that is more applicable to science, the literature has employed a different definition of "probability function": Lattice additivity is altered so that it only applies to certain pairs of disjoint lattice elements. This version of "probability function" has proved useful in quantum physics and is discussed in Chapter 6. Chapter 4 pursues approaches to probability theory using distributive lattices, including situations with lattices that are complemented but not pseudo complemented (generalization (i) above) as well as lattices without complementation operations. The chapter also discusses the relationship of these approaches to rationality considerations.

Birkhoff & von Neumann (1936) generalized probability theory by using complemented modular lattices as event spaces. The following theorem shows that multiple complementations must be assumed for such lattices to achieve a proper generalization.

Theorem 2.22 (Birkhoff - von Neumann Theorem). *Each uniquely complemented modular lattice is boolean.*

   Proof. Let $\mathfrak{L} = \langle L, \sqcup, \sqcap, \neg, 1, 0 \rangle$ be a uniquely complemented modular lattice. By Theorem 2.19, it only needs to be shown that $\neg$ is a pseudo complementation operation. Let a and b be arbitrary elements of L such that

$$a \sqcap b = 0.$$

Because

$$a \le a \sqcup \neg (a \sqcup b),$$

it is sufficient to show that

$$\neg b = a \sqcup \neg (a \sqcup b). \tag{2.52}$$

Using the fact that $a \le a \sqcup b$ and the $\sqcup$-modular law (Theorem 2.10), it follows that

$$a = a \sqcup 0 = a \sqcup [(a \sqcup b) \sqcap \neg (a \sqcup b)] = (a \sqcup b) \sqcap [a \sqcup \neg (a \sqcup b)].$$

This implies that

$$0 = a \sqcap b = [a \sqcup \ulcorner (a \sqcup b)] \sqcap [(a \sqcup b) \sqcap b] = [a \sqcup \ulcorner (a \sqcup b)] \sqcap b. \quad (2.53)$$

Because

$$1 = (a \sqcup b) \sqcup \ulcorner (a \sqcup b) = [a \sqcup \ulcorner (a \sqcup b)] \sqcup b, \quad (2.54)$$

it follows from the uniqueness of $\ulcorner$ as the complement operation on $\mathfrak{L}$ and Equations 2.53 and 2.54 that

$$\ulcorner b = a \sqcup \ulcorner (a \sqcup b),$$

showing Equation 2.52. $\quad \square$

Salii (1984) remarks that Theorem 2.22 has been associated with Skolem, Bergmann, and Huntington, but "The first to give a proof (albeit a very wordy one) was Birkhoff (1948), who alluded to a more general result of von Neumann." *(Salii, 1984, p. 41)*

2.9   Orthomodular Lattices

This section provides some basic properties of orthomodular lattices. Orthomodular lattices are used in physics to describe the algebra of events involving quantum phenomena. They retain many properties of boolean lattices, except possibly for distributivity and unique complementation. Like boolean lattices, they have a complementation operation that satisfies DeMorgan's Laws.

Definition 2.24 (ortholattice). A complemented lattice satisfying De-Morgan's Laws is called an *ortholattice*. A complementation operation of an ortholattice that satisfies DeMorgan's Laws is often denoted by the symbol $\perp$. $\quad \square$

Definition 2.25 (orthomodular). $\mathfrak{L} = \langle L, \sqcup, \sqcap, \perp, 1, 0 \rangle$ is said to be an *orthomodular lattice* if and only if $\mathfrak{L}$ is an ortholattice and the following condition holds:

Orthomodularity: For all x and y in L, if $x \leq y$ then $x \sqcup (x^\perp \sqcap y) = y$. $\quad \square$
$$(2.55)$$

It is convenient for some proofs to have Equation 2.55 stated in a slightly different form.

Lemma 2.9. *Let* $\mathfrak{L} = \langle L, \sqcup, \sqcap, {}^{\perp}, 1, 0 \rangle$ *be an ortholattice. Then* $\mathfrak{L}$ *is orthomodular (Equation 2.55) if and only if it satisfies the following equation:*

$$\text{For all } x \text{ and } y \text{ in } L, \ x \sqcup (x^{\perp} \sqcap (x \sqcup y)) = x \sqcup y. \qquad (2.56)$$

Proof. Suppose $\mathfrak{L}$ is orthomodular. Let x and y be arbitrary elements of L. Then $x \leq x \sqcup y$. Thus, by substituting $(x \sqcup y)$ in place of y in Equation 2.55,

$$x \sqcup [x^{\perp} \sqcap (x \sqcup y)] = x \sqcup y,$$

showing Equation 2.56.

Suppose Equation 2.56. Suppose $x \leq y$. Then $y = x \sqcup y$. Thus by Equation 2.56,

$$x \sqcup (x^{\perp} \sqcap y) = y,$$

showing Equation 2.55. □

The following theorem shows that all modular lattices are orthomodular.

Theorem 2.23. *Let* $\mathfrak{L} = \langle L, \sqcup, \sqcap, {}^{\perp}, 1, 0 \rangle$ *be a modular ortholattice. Then* $\mathfrak{L}$ *is orthomodular.*

Proof. By Theorem 2.10 $\mathfrak{L}$ is $\sqcup$-modular. Let x and y be arbitrary elements of L such that $x \leq y$. Then, by $\sqcup$-modularity,

$$(x \sqcup y) \sqcap (x \sqcup x^{\perp}) = x \sqcup (y \sqcap x^{\perp}),$$

which, because $x \leq y$, yields

$$y = x \sqcup (y \sqcap x^{\perp}),$$

showing orthomodularity (Equation 2.56). □

Theorem 2.24. *All boolean lattices are orthomodular.*

Proof. Let $\mathfrak{B}$ be a boolean lattice. Then by Theorem 2.15, $\mathfrak{B}$ is an ortholattice. Because $\mathfrak{B}$ is boolean, it is distributive, and therefore, by the definitions of "modular" and "distributive", it is modular. Thus, by Theorem 2.23, $\mathfrak{B}$ is orthomodular. □

The following concept plays an important role later in the development of orthoprobability theory.

Definition 2.26 (x commutes with y, xCy). For elements x and y of an ortholattice, x is said to *commute* with y, in symbols, xCy, if and only if $x = (x \sqcap y) \sqcup (x \sqcap y^{\perp})$. □

Note that $xC0$ and $1Cx$ always hold and $x \leq y$ implies xCy. The use of the word "commute" in Definition 2.26 comes from a special use of it in the study of projection operators in hilbert space. For ortholattices, it is not the algebraic use of the word "commute", because from xCy it does not follow that yCx. However, for orthomodular lattices it does correspond to the algebraic use of "commute". (See Statements (1) and (5) of Theorem 2.25 below.)

The following lattice also plays an important role later in the development of orthoprobability theory.

Definition 2.27 (lattice O_6). By definition, each *lattice O_6* is a six-element ortholattice, $0, 1, x, x^{\perp}, y, y^{\perp}$, where $x^{\perp}$ and $y^{\perp}$ are, respectively, the orthocomplements of x and y, and where $0 < x < y < 1$ and $0 < y^{\perp} < x^{\perp} < 1$. By inspection, a O_6 lattice exists and is described in Figure 2.4. □

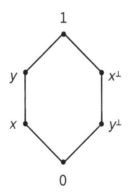

Fig. 2.4   Lattice O_6

Definition 2.28 (subalgebra, ΓA). Let $\mathfrak{L} = \langle L, \sqcup, \sqcap, \frown, 1, 0 \rangle$ with a negative operator $\frown$. A *subalgebra* of $\mathfrak{L}$ is a subset S such that $0 \in S$, $1 \in S$, and for all x and y in S, $x \sqcup y$, $x \sqcap y$, and $\frown x$ are in S. If A is a subset of L, then ΓA is, by definition, the smallest subalgebra of $\mathfrak{L}$ containing the set A. It easily follows that ΓA exists for each subset A of L. □

Note the following difference in "subalgebra of $\mathfrak{L}$" and "sublattice of $\mathfrak{L}$": A subalgebra of $\mathfrak{L}$ is a sublattice of $\mathfrak{L}$ that is required to have the unit element 1 and zero element 0 of $\mathfrak{L}$ as elements of the subalgebra, whereas a sublattice of $\mathfrak{L}$ is not required to have these elements (unless $\mathfrak{L}$ is the lattice consisting of 0 and 1).

Theorem 2.25. *(Theorem 2, pp. 22-23, of Kalmbach, 1983) Let $\mathfrak{L} = \langle L, \sqcup, \sqcap, \perp, 1, 0 \rangle$ be an ortholattice. Then the following five statements are equivalent:*

(1) $\mathfrak{L}$ is orthomodular.
(2) For all x and y in L, if $x \leq y$ and $y \sqcap x^\perp = 0$, then $x = y$.
(3) O_6 is not a subalgebra of L.
(4) For all x and y in L, if $x \leq y$, then $\Gamma\{x, y\}$ is a boolean subalgebra of $\mathfrak{L}$.
(5) For all x and y in L, xCy iff yCx.

Proof. Suppose Statement (1) that $\mathfrak{L}$ is orthomodular. Suppose x and y are arbitrary elements in L, $x \leq y$ and $y \sqcap x^\perp = 0$. Then by orthomodular law, $x = x \sqcup (y \sqcap x^\perp) = y$, showing Statement (2).

Suppose Statement (2) that if $x \leq y$ and $y \sqcap x^\perp = 0$ then $x = y$. Let S be an arbitrary O_6 lattice of $\mathfrak{L}$, say described by Figure 2.4. Then $x \leq y$ and $y \sqcap x^\perp = 0$, but $x \neq y$, contradicting that S is an O_6 lattice. This shows Statement (3).

Suppose Statement (3) that O_6 is not a subalgebra of $\mathfrak{L}$. Assume $x \leq y$. Suppose $\Gamma\{x, y\}$ is not a boolean lattice. To show Statement (4), only a contradiction needs to be shown. Because $\mathfrak{L}$ is an ortholattice, it then follows that $\Gamma\{x, y\}$ is not distributive. Also because $\mathfrak{L}$ is an ortholattice, it follows from Theorem 2.14 that $y^\perp \leq x^\perp$ and $y^{\perp\perp} = y$. The following will be shown:

$$\text{either } x \sqcup (x^\perp \sqcap y) \neq y \text{ or } y^\perp \sqcup (x^\perp \sqcap y) \neq x^\perp. \qquad (2.57)$$

Suppose Equation 2.57 did not hold. Then

$$x \sqcup (x^\perp \sqcap y) = y \text{ and } y^\perp \sqcup (x^\perp \sqcap y) = x^\perp. \qquad (2.58)$$

It then follows that

$$\Gamma\{x, y\} = \{0, 1, x, x^\perp, y, y^\perp, x^\perp \sqcap y, x \sqcup y^\perp\}$$

and that $\Gamma\{x, y\}$ is distributive. This is contrary to the hypothesis that $\Gamma\{x, y\}$ is not distributive. Therefore Equation 2.57 must hold. Without loss of generality, suppose

$$x \sqcup (x^\perp \sqcap y) \neq y.$$

Then the subalgebra,

$$0, 1, x \sqcup (x^\perp \sqcap y), y, y^\perp, x^\perp \sqcap (x \sqcup y^\perp)$$

is an 0_6 subalgebra of $\mathfrak{L}$, contradicting Statement (3).

Suppose Statement (4) that for all u and v in L, if $u \leq v$ then $\Gamma\{u, v\}$ is a boolean subalgebra of $\mathfrak{L}$. It is sufficient to show that xCy implies yCx. Suppose xCy. Then

$$x = (x \sqcap y) \sqcup (x \sqcap y^\perp),$$

and thus,

$$x^\perp = (x^\perp \sqcup y^\perp) \sqcap (x^\perp \sqcup y).$$

Therefore,

$$x^\perp \sqcap y = [(x^\perp \sqcup y^\perp) \sqcap (x^\perp \sqcup y)] \sqcap y = (x^\perp \sqcup y^\perp) \sqcap y = (x \sqcap y)^\perp \sqcap y. \quad (2.59)$$

Because $x \sqcap y \leq y$, it follows from Statement (4) that $\Gamma\{x \sqcap y, y\}$ is boolean and therefore distributive. Thus, by Equation 2.59 and $x \sqcap y \leq y$,

$$(x \sqcap y) \sqcup (x^\perp \sqcap y) = (x \sqcap y) \sqcup [(x \sqcap y)^\perp \sqcap y)] = 1 \sqcap [(x \sqcap y) \sqcup y] = y,$$

showing yCx and therefore Statement (5).

Suppose Statement (5) that xCy iff yCx. Suppose $x \leq y$. Then

$$x = (x \sqcap y) \sqcup (x \sqcap y^\perp),$$

showing xCy. By Statement (5), yCx, that is,

$$y = (y \sqcap x) \sqcup (y \sqcap x^\perp) = x \sqcup (x^\perp \sqcap y),$$

which shows orthomodularity, Statement (1). $\quad \square$

Definition 2.29 (atom, atomic, atomistic). Let $\mathfrak{L} = \langle L, \leq, \sqcup, \sqcap, 1, 0 \rangle$ be a lattice. Then the following definitions hold:

- An element a of L is said to be an *atom (of L)* if and only if $0 < a$ and for all b in L, if $0 \leq b \leq a$ then $b = 0$ or $b = a$.
- $\mathfrak{L}$ is said to be *atomic* if and only if for each b in L, if $b \neq 0$ then there exists an atom a of $\mathfrak{L}$ such that $a \leq b$.
- $\mathfrak{L}$ is said to be *atomistic* if and only if for each $b \neq 0$ in L,

  b is the $\leq$-least upper bound of $\{a \mid a$ is an atom and $a \leq b\}$. $\quad \square$

Theorem 2.26. *Let* $\mathfrak{L} = \langle L, \leq, \sqcup, \sqcap, 1, 0 \rangle$ *be a finite lattice. Then* $\mathfrak{L}$ *is atomic.*

Proof. Because $\mathfrak{L}$ is finite, let n be a positive integer such that L has n elements. Assume $\mathfrak{L}$ is not atomic. A contradiction will be shown. Because $\mathfrak{L}$ is not atomic, let $a_1 \neq 0$ be an element of L such that there does not exist an atom a of L such that $a \leq a_1$. Since a_1 is not an atom, an element a_2 of L can be found such that a_2 is not an atom and $0 < a_2 < a_1$. This process can be repeated until elements $a_1, \ldots, a_n$ of L are found so that

$$0 < a_n < \cdots < a_1 ,$$

contradicting that L has n elements. $\square$

Theorem 2.27. *Let* $\mathfrak{L} = \langle L, \leq, \sqcup, \sqcap, 1, 0 \rangle$ *be an atomic orthomodular lattice. The* $\mathfrak{L}$ *is atomistic.*

Proof. Let b be an arbitrary element of L such that $b \neq 0$, and let

$$\mathcal{A} = \{a \mid a \text{ is an atom of } L \text{ and } a \leq b\} .$$

Then, because $\mathfrak{L}$ is atomic and $b \neq 0$, it follows that $\mathcal{A} \neq \varnothing$. It needs to be shown that b is the $\leq$-least upper bound of $\mathcal{A}$. Let c be an arbitrary element of L such that $a \leq c$ for each a in $\mathcal{A}$. Because $b \sqcap c \leq b$, it follows by orthomodularity that

$$(b \sqcap c) \sqcup (b \sqcap (b \sqcap c)^{\perp}) = b . \tag{2.60}$$

There are two cases to consider.

Case 1: $b \sqcap (b \sqcap c)^{\perp} \neq 0$. Because $\mathfrak{L}$ is atomic, let d be an atom of L such that

$$d \leq b \sqcap (b \sqcap c)^{\perp} .$$

Then $d \leq b$ and $d \leq (b \sqcap c)^{\perp}$. It follows from $d \leq b$ that $d \in \mathcal{A}$ and thus $d \leq c$, that is, $d \leq (b \sqcap c)$. This contradicts $d \leq (b \sqcap c)^{\perp}$.

Case 2: $b \sqcap (b \sqcap c)^{\perp} = 0$. Then by Equation 2.60, $b \sqcap c = b$, and thus $b \leq c$. $\square$

Theorem 2.28. *Let* $\mathfrak{L} = \langle L, \leq, \sqcup, \sqcap, 1, 0 \rangle$ *be a finite orthomodular lattice. Then for each* $b \neq 0$ *in* L, *there is a finite set of atoms of* L, $a_1, \ldots, a_n$ *such that* $b = a_1 \sqcup \cdots \sqcup a_n$.

Proof. Let b be an arbitrary element of L. Then by Theorem 2.26, $\mathfrak{L}$ is atomic, and by Theorem 2.27, it is atomistic. Thus let $a_1, \ldots, a_n$ be elements of L such that b is the $\leq$-least upper bound of $\{a_1, \ldots, a_n\}$. Then, because $x \sqcup y$ is the $\leq$-least upper bound for all x and y in L, it easily follows that

$$a_1 \sqcup \cdots \sqcup a_n \leq b ,$$

which by the choice of b implies $b = a_1 \sqcup \cdots \sqcup a_n$. $\square$

2.10 Finite Boolean Lattices

Theorem 2.29. *Suppose $\mathfrak{L} = \langle L, \leq, \sqcup, \sqcap, 1, 0 \rangle$ is a finite boolean lattice and $b \neq 0$ is an arbitrary element of L. Then there exists a unique set of atoms of L, $\{a_1, \ldots, a_n\}$ such that*

$$b = a_1 \sqcup \cdots \sqcup a_n.$$

Proof. Because $\mathfrak{L}$ is boolean, it is an orthomodular lattice by Theorem 2.24. Thus, by Theorem 2.28, let $a_1, \ldots, a_n$ be atoms of L such that $b = a_1 \sqcup \cdots \sqcup a_n$. Suppose $c_1, \ldots, c_m$ are atoms of L such that also $b = c_1 \sqcup \cdots \sqcup c_m$. Without loss of generality, suppose $a_1, \ldots, a_n$ are distinct and $c_1, \ldots, c_n$ are distinct. Then to show the theorem, it needs only to be shown that

$$\{a_1, \ldots, a_n\} = \{c_1, \ldots, c_m\}.$$

Note that if x and y are atoms of L, then

- if $x \neq y$ then $x \sqcap y = \varnothing$, and
- if $x = y$ then $x \sqcap y = x$.

Because $b = a_1 \sqcup \cdots \sqcup a_n = c_1 \sqcup \cdots \sqcup c_m$ and $\mathfrak{L}$ is distributive, it follows that $a_1 \leq b$ and thus that

$$a_1 \sqcap (c_1 \sqcup \cdots \sqcup c_m) = (a_1 \sqcap c_1) \sqcup \cdots \sqcup (a_1 \sqcap c_m) = a_1.$$

Because a_1 and $c_1, \ldots, c_m$ are atoms of L and $c_1, \ldots, c_m$ are distinct, this can only happen if exactly one of the c_i is a_1. Similar arguments hold for $a_2, \ldots, a_n$. Thus

$$\{a_1, \ldots, a_n\} \subseteq \{c_1, \ldots, c_m\}.$$

By a similar argument, $\{c_1, \ldots, c_m\} \subseteq \{a_1, \ldots, a_n\}$. Therefore,

$$\{a_1, \ldots, a_n\} = \{c_1, \ldots, c_m\}. \quad \square$$

Theorem 2.30. *Suppose $\mathfrak{Z} = \langle \mathcal{Z}, \cup, \cap, -, Z, \varnothing \rangle$ is a finite boolean algebra of events. Then there exist a finite subset $\mathcal{X}$ of $\mathcal{Z}$ such that $\mathfrak{Z} = \langle \wp(\mathcal{X}), \cup, \cap, -, Z, \varnothing \rangle$.*

Proof. Let $\mathcal{X}$ be the set of atoms of $\mathcal{Z}$. Then for each nonempty subset $\mathcal{C}$ of $\mathcal{X}$, $\bigcup \mathcal{C}$ is in $\mathcal{Z}$, and by Theorem 2.29, each element of $\mathcal{Z}$ is such a finite union of elements of $\mathcal{X}$. In particular, $Z = \bigcup \mathcal{X}$. By construction, $\mathfrak{Z} = \langle \wp(\mathcal{X}), \cup, \cap, -, X, \varnothing \rangle. \quad \square$

Definition 2.30 (finitely generated boolean subalgebra). Let $\mathfrak{L} = \langle L, \sqcup, \sqcap, -, 1, 0 \rangle$ be a boolean lattice and S a nonempty subset of L. Then $\mathfrak{B} = \langle B, \sqcup, \sqcap, -, 0, 1 \rangle$ is said to be the *boolean subalgebra generated by S* if and only if B is the smallest subset of L such that $S \subseteq B$ and for all a and b in B, $(a \sqcup b) \in B$, $(a \sqcap b) \in B$, and $-a \in B$. It follows from this definition of $\mathfrak{B}$ that it exists and is a subalgebra of $\mathfrak{L}$. It is said to be *finitely generated* if and only if S is a finite set. $\square$

Theorem 2.31. *Suppose S is a nonempty finite subset of a boolean lattice $\mathfrak{L} = \langle L, \sqcup, \sqcap, -, 1, 0 \rangle$. Then the boolean lattice generated by S,*

$$\mathfrak{B} = \langle B, \sqcup, \sqcap, -, 1, 0 \rangle \,,$$

is a finite boolean subalgebra of $\mathfrak{L}$.

Proof by induction. If S has 1 element, say a, then $B = \{1, 0, a, -a\}$ and the theorem has been shown.

Suppose $n - 1 \geq 1$, L has more than $n - 1$ elements, and T consists of the $n - 1$ elements $t_1, \ldots, t_{n-1}$ of L. By inductive hypothesis, let $\mathfrak{F} = \langle F, \sqcup, \sqcap, 1, 0 \rangle$ be the finite boolean subalgebra of $\mathfrak{L}$ generated by T. Let $a_1, \ldots, a_m$ be a listing of the elements of F, and let a be an arbitrary element of $L - T$. To complete the induction, it needs to only show that $S = T \cup \{a\}$ generates a finite boolean algebra of $\mathfrak{L}$.

Let

$$G = \{a \sqcap f \mid f \in F\} \cup \{(-a) \sqcap f \mid f \in F\} \,.$$

Then, because F is finite, G is finite. Because $1 \in F$ and $a \sqcap 1 = a$, it follows that

$$a \in G \,. \tag{2.61}$$

Then, because $\mathfrak{F}$ is a boolean subalgebra of $\mathfrak{L}$, $a \sqcap 1 = a$ and $(-a) \sqcap 1 = -a$, and therefore, $a \sqcup (-a) = 1$ and therefore,

$$1 \in G \,. \tag{2.62}$$

Also, because $a \sqcap 0 = 0$,

$$0 \in G \,. \tag{2.63}$$

Let $g = x \sqcap b$ and $h = y \sqcap c$ be arbitrary elements of G, where $x = a$ or $x = (-a)$, and $y = a$ or $y = (-a)$, and where b and c are in F. If $x = y$, then

$$g \sqcap h = (x \sqcap b) \sqcap (x \sqcap c) = x \sqcap (b \sqcap c)$$

is in G, because $b \sqcap c$ is in F. If $x \neq y$, then

$$g \sqcap h = (x \sqcap b) \sqcap (y \sqcap c) = (x \sqcap y) \sqcap (b \sqcap c) = 0 \sqcap (b \sqcap c) = 0,$$

and thus $(g \sqcap h) \in G$. The above shows that

$$(g \sqcap h) \in G. \tag{2.64}$$

It will next be shown that $-g \in G$. This will be done by describing an element k of F that is the complement of g. Let

$$k = (x \sqcap -b) \sqcup (-x \sqcap 1).$$

Then

$$k \in G. \tag{2.65}$$

By distributivity in $\mathfrak{L}$,

$$\begin{aligned}
g \sqcap k &= (x \sqcap b) \sqcap [(x \sqcap -b) \sqcup (-x \sqcap 1)] \\
&= [(x \sqcap b) \sqcap (x \sqcap -b)] \sqcup [(x \sqcap b) \sqcap (-x \sqcap 1)] \\
&= 0 \sqcup [(x \sqcap b) \sqcap -x] \\
&= 0,
\end{aligned}$$

and

$$\begin{aligned}
g \sqcup k &= (x \sqcap b) \sqcup [(x \sqcap -b) \sqcup (-x \sqcap 1)] = [(x \sqcap b) \sqcup (x \sqcap -b)] \sqcup (-x \sqcap 1) \\
&= [x \sqcap (b \sqcup -b)] \sqcup -x = x \sqcup -x = 1.
\end{aligned}$$

The above establishes that k is in G and $-g = k$. Thus

$$-g \text{ is in } G. \tag{2.66}$$

Because $\mathfrak{L}$ is boolean and g and h are in L,

$$g \sqcup h = (--g) \sqcup (--h) = -[(-g) \sqcap (-h)],$$

and thus by Equations 2.64 and 2.66,

$$g \cup h \text{ is in } G. \tag{2.67}$$

Equations 2.61 to 2.67 with the fact that G is a subset of elements from the distributive lattice $\mathfrak{L}$ show that $\langle G, \cup, \cap, -, 1, 0 \rangle$ is a boolean lattice. $\square$

An immediate consequence of Theorem 2.31 is an analogous result for distributive lattices:

Definition 2.31. Suppose S is a nonempty finite subset of a distributive lattice $\mathfrak{L} = \langle L, \sqcup, \sqcap, 1, 0 \rangle$. Then the *distributive lattice generated by S* is the smallest subset D of L such that $S \subseteq D$, $1 \in D$, $0 \in D$, and for all a and b in D, $a \sqcup b$ is in D and $a \sqcap b$ is in D. $\square$

Theorem 2.32. *Suppose S is a nonempty finite subset of a distributive lattice $\mathfrak{L}$. Then the distributive lattice generated by S is a finite distributive subalgebra of $\mathfrak{L}$.*

Proof. By Theorem 2.31, the boolean lattice $\mathfrak{B}$ generated by S is finite, and obviously the distributive lattice generated by S is a subalgebra of $\mathfrak{B}$ and therefore is finite. $\square$

Chapter 3

Pseudo Complemented Distributive Lattices

This chapter describes how distributive and pseudo complemented distributed lattices capture the probabilistic algebra inherent in traditonal probability theory. The mathematician Brouwer employed a special case of them, today called "intuitionistic propositional logic", to provide a logic for his radical foundation of mathematics. Others later found that Brouwer's logic was useful in various issues outside of his foundation for mathematics. Today, intuitionistic propositional logic is an important tool in philosophy, mathematics, and computer science. For this chapter, however, it is its generalizations to lattices that describe probabilistic event spaces that are of primary interest.

Pseudo complemented distributive lattices come in various kinds. Boolean lattices turn out to be the special case where pseudo complementation is complementation. Between general pseudo complemented distributive lattices and boolean lattices are heyting, stone, and topological lattices:

- Heyting lattices are pseudo complemented distributive lattices that have an additional binary operation corresponding to the logical implication connective of intuitionistic propositional logic.
- A stone lattice is a pseudo complemented distributive lattice such that the set of pseudo complemented elements form a boolean subalgebra. Stone lattices are used in computer science and engineering as (i) deterministic alternatives to fuzzy set theory, and (ii) as parts of methodologies for reducing the complexity of event algebras.
- Topological lattices consists of the open sets from a topological space together with $\cup$ as join, $\cap$ as meet, and an operation of pseudo complementation, $\dot{-}$.

A theorem presented later in this chapter shows that each pseudo complemented distributive lattice is isomorphic to a sublattice algebra of a topological lattice. This result shows that for probabilistic considerations, boolean lattices, pseudo complemented distributive lattices, heyting lattices, and stone lattices can be viewed as topological lattices of events. For such topological representations, complementation and pseudo complementation have simple and useful topological interpretations. Chapter 5 exploits these topological interpretations to formulate concepts of ambiguity, vagueness, incompleteness that are useful in decision theory.

3.1 Pseudo Complementation and Relative Pseudo Complementation

This section generalizes boolean lattices by keeping distributivity but generalizing complementation to pseudo complementation and to relative pseudo complementation (Definitions 3.1 and 3.3 below). These generalizations are much discussed in the philosophy of mathematics literature. They correspond to the negation and implication connectives of intuitionistic logic. The lattice version of intuitionistic negation is called "pseudo complementation," and has been discussed in Chapters 1 and 2. The lattice version of intuitionistic implication is called "relative pseudo complementation." Interestingly, relative pseudo complementation has strong enough uniqueness properties built into its definition so that distributivity follows from its existence (Theorem 3.3 below).

Definition 3.1 (pseudo complementation). Let $\mathfrak{L} = \langle L, \leq, \sqcup, \sqcap, 1, 0 \rangle$ be a lattice and a, b, and c be arbitrary elements of L. Then the following definitions hold:

- c is said to be the *pseudo complement of* a if and only if for each x in L,

$$x \leq c \text{ iff } a \sqcap x = 0 \,.$$

- $\mathfrak{L}$ is said to be *pseudo complemented* if and only if each element of L is pseudo complemented. □

The following theorem, which follows immediately from the definition of "pseudo complement", establishes the uniqueness of pseudo complementation.

Theorem 3.1. *Suppose* $\mathfrak{L} = \langle L, \leq, \sqcup, \sqcap, 1, 0 \rangle$ *is a lattice. Then for each element* a *of* L, *if the pseudo complement of* a *exists, then it is unique.*
    Proof. Immediate from Definition 3.1.   □

Theorem 3.1 justifies the following convention.

Convention 3.1. *A pseudo complemented lattice,* $\mathfrak{L} = \langle L, \leq, \sqcup, \sqcap, 1, 0 \rangle$, *is often written as* $\mathfrak{L} = \langle L, \leq, \sqcup, \sqcap, \neg, 1, 0 \rangle$, *with the understanding that* $\neg$ *is the operation of pseudo complementation on* $\mathfrak{L}$.   □

Definition 3.2. pseudo complemented distributive lattice. $\mathfrak{L}$ is said to be a *pseudo complemented distributive lattice* if and only if it is a pseudo complemented, distributive lattice.   □

The following definition provides a formal definition of "relative pseudo complementation". It does not, however, provide intuitive insight as to why it is an important concept. Theorems presented later in this chapter describe its relationship to the implication connective of classical logic.

Definition 3.3 (relative pseudo complementation). Let

$$\mathfrak{L} = \langle L, \leq, \sqcup, \sqcap, 1, 0 \rangle$$

be a lattice and a, b, and c be arbitrary elements of L. Then the following definitions hold:

- c is said to be a *pseudo complement of* a *relative to* b, in symbols, $c = (a \Rightarrow b)$, if and only if for all x in L ,

$$x \leq c \ \text{iff} \ a \sqcap x \leq b. \ \ \square$$

- $\mathfrak{L}$ is said to be *relatively pseudo complemented* if and only if for all a and b in L, $a \Rightarrow b$ exists. If $\mathfrak{L}$ is relatively pseudo complemented, the operation $\Rightarrow$ is called the *operation of relative pseudo complementation* for $\mathfrak{L}$.
- $\mathfrak{L}$ is said to be a *heyting lattice* if and only if it is relatively pseudo complemented.   □

The following theorem establishes the uniqueness of the relative pseudo complementation operator.

Theorem 3.2. *Suppose* $\mathfrak{L} = \langle L, \leq, \sqcup, \sqcap, 1, 0 \rangle$ *is a lattice and* a *and* b *are elements of* L. *Then* $a \Rightarrow b$ *is a unique element of* L.
    Proof. Immediate from Definition 3.3.   □

Theorem 3.2 justifies the following convention.

Convention 3.2. Throughout this book, unless otherwise specified, $\Rightarrow$ denotes the operation of relative pseudo complementation of a relative pseudo complemented lattice. Relative pseudo complemented lattices,

$$\mathfrak{L} = \langle L, \leq, \sqcup, \sqcap, 1, 0 \rangle \,,$$

are often written as

$$\mathfrak{L} = \langle L, \leq, \sqcup, \sqcap, \Rightarrow, 1, 0 \rangle \,,$$

with the understanding that $\Rightarrow$ is the operation of relative pseudo complementation on $\mathfrak{L}$. $\square$

The following theorem shows that for each element a of a heyting lattice, the pseudo complement of a, $\ulcorner a$, exists and has a definition as $a \Rightarrow 0$. Example 3.1 below shows that $\ulcorner$ can exist in a distributive lattice without $\Rightarrow$ existing in the same lattice, that is, there are pseudo complemented distributive lattices that are not heyting lattices. Example 3.2 shows that there is a heyting lattice $\mathfrak{L}$ such that $\Rightarrow$ cannot be defined algebraically in terms of $\mathfrak{L}$'s operations $\sqcap$, $\sqcup$, and $\ulcorner$.

In a heyting lattice, $\Rightarrow$ has the following important property of an "implication operator": If $a = 1$ (i.e., a is true) and $(a \Rightarrow b) = 1$ (i.e., $a \Rightarrow b$ is true) then $b = 1$ (i.e., b is true).

Theorem 3.3. *Suppose* $\mathfrak{L} = \langle L, \leq, \sqcup, \sqcap, \ulcorner, \Rightarrow, 1, 0 \rangle$ *is a heyting lattice. Then* $\mathfrak{L}$ *is a pseudo complemented distributive lattice and for each* a *in* L, *the pseudo complement of* a, $\ulcorner a$, *is defined by,*

$$\ulcorner a \;=\; a \Rightarrow 0 \,. \tag{3.1}$$

Proof. It is immediate from the definitions of "relative pseudo complement", "pseudo complement", and Equation 3.1 that $\mathfrak{L}$ is pseudo complemented, with $\ulcorner$ being its pseudo complement. Thus to show $\mathfrak{L}$ is a pseudo complemented distributive lattice, it needs to show that $\mathfrak{L}$ is distributive. Let a, b, and c be arbitrary elements of L. Because

$$a \sqcap b \leq (a \sqcap b) \sqcup (a \sqcap c) \text{ and } a \sqcap c \leq (a \sqcap b) \sqcup (a \sqcap c) \,, \tag{3.2}$$

it follows from Definition 3.3 that

$$b \;\leq\; a \Rightarrow [(a \sqcap b) \sqcup (a \sqcap c)] \text{ and } c \;\leq\; a \Rightarrow [(a \sqcap b) \sqcup (a \sqcap c)] \,.$$

Thus,

$$b \sqcup c \;\leq\; a \Rightarrow (a \sqcap b) \sqcup (a \sqcap c) \,,$$

and therefore, by Definition 3.3,

$$a \sqcap (b \sqcup c) \leq (a \sqcap b) \sqcup (a \sqcap c). \tag{3.3}$$

However, because,

$$a \sqcap b \leq a \sqcap (b \sqcup c) \text{ and } a \sqcap c \leq a \sqcap (b \sqcup c),$$

it follows that

$$(a \sqcap b) \sqcup (a \sqcap c) \leq a \sqcap (b \sqcup c). \tag{3.4}$$

By Equations 3.3 and 3.4,

$$a \sqcap (b \sqcup c) = (a \sqcap b) \sqcup (a \sqcap c). \quad \square$$

The implication operator of classical propositional logic, $\rightarrow$, is defined algebraically in terms of disjunction, $\vee$, conjunction, $\wedge$, and negation, $\neg$, as follows: For all propositions α and β,

$$\alpha \rightarrow \beta \text{ is by definition } (\neg \alpha) \vee \beta.$$

In boolean lattices, this takes the form: For all elements a and b of the lattice,

$$a \rightarrow b = (-a) \sqcup b.$$

The equation,

$$a \Rightarrow b = (\ulcorner a) \sqcup b,$$

can fail for some elements a and b of a heyting lattice that is not a boolean lattice.

Example 3.1. An example of a pseudo complemented distributive lattice that is not a heyting lattice. Let

$$X = \{a, 1, 2, 3, \ldots\},$$

$$\mathcal{X} = \{\varnothing, X,$$
$$\{1\}, \{1, 2\}, \{1, 2, 3\}, \{1, 2, 3, 4\}, \ldots,$$
$$\{a, 1\}, \{a, 1, 2\}, \{a, 1, 2, 3\}, \{a, 1, 2, 3, 4\}, \ldots\},$$

and

$$\ulcorner \varnothing = X, \ \ulcorner X = \varnothing,$$
$$\varnothing = \ulcorner \{1\} = \ulcorner \{1, 2\} = \ulcorner \{1, 2, 3\} = \ulcorner \{1, 2, 3, 4\} = \cdots,$$
$$\varnothing = \ulcorner \{a, 1\} = \ulcorner \{a, 1, 2\} = \ulcorner \{a, 1, 2, 3\} = \ulcorner \{a, 1, 2, 3, 4\} = \cdots.$$

Then, by inspection, $\mathfrak{X} = \langle \mathcal{X}, \cup, \cap, \llcorner, X, \varnothing \rangle$ is a pseudo complemented distributive lattice. However, $\mathfrak{X}$ is not a heyting lattice, because $\{a, 1\} \Rightarrow \{1\}$ does not exist. This is because $a \notin (\{a, 1\} \Rightarrow \{1\})$ and thus $X \neq (\{a, 1\} \Rightarrow \{1\})$, but

$$1 \in (\{a, 1\} \Rightarrow \{1\}), \; 2 \in (\{a, 1\} \Rightarrow \{1\}), \; 3 \in (\{a, 1\} \Rightarrow \{1\}), \; \ldots \, .$$

But this implies that

$$\{a, 1\} \Rightarrow \{1\} \; = \; \{1, 2, 3, 4, \ldots\} \, ,$$

which implies that $\{1, 2, 3, 4, \ldots\} \in \mathcal{X}$, which cannot be the case by the definition of $\mathcal{X}$. $\quad \Box$

Example 3.2. An example of a heyting lattice where $\Rightarrow$ is not definable algebraically in terms of $\cup$, $\cap$, and $\llcorner$. (The construction used in this example comes from Schechter, 2005.) Let

$$X = \{1, 2, 3, 4\} \, ,$$

$$\mathcal{X} = \{\varnothing, \{1\}, \{2\}, \{1, 2\}, \{1, 3\}, \{2, 4\}, \{1, 2, 3\}, \{1, 2, 4\}, X\},$$

and

$$\llcorner \varnothing = X, \; \llcorner \{1\} = \{2, 4\}, \; \llcorner \{2\} = \{1, 3\}, \; \llcorner \{1, 2\} = \varnothing, \; \llcorner \{1, 3\} = \{2, 4\},$$

$$\llcorner \{2, 4\} = \{1, 3\}, \; \llcorner \{1, 2, 3\} = \varnothing, \; \llcorner \{1, 2, 4\} = \varnothing, \; \llcorner X = \varnothing \, .$$

Then, by inspection, $\mathfrak{L} = \langle X, \cup, \cap, \llcorner, X, \varnothing \rangle$ is a heyting lattice. Furthermore, $(\{1, 2, 3\} \Rightarrow \{1, 2\}) = \{1, 2, 4\}$. However,

$$Y = \{\varnothing, \; \{1, 2\}, \; \{1, 2, 3\}, \; X\}$$

is closed under the operations $\cup$, $\cap$, and $\llcorner$, and therefore $\Rightarrow$ cannot be defined algebraically in terms of $\cup$, $\cap$, and $\llcorner$, because such a definition when applied to $\{1, 2, 3\} \Rightarrow \{1, 2\}$ produces the event $\{1, 2, 4\}$ that has an element, 4, that is outside of Y. $\quad \Box$

It easily follows from the definition of "heyting" that all finite distributive lattices are heyting. Thus, examples of pseudo complemented distributive lattices that are not heyting must be infinite.

Theorems 3.4 and 3.5 below describe some important algebraic features of distributive lattices involving complementation and pseudo complementation.

Theorem 3.4. *Suppose $\mathfrak{L} = \langle L, \leq, \sqcup, \sqcap, \llcorner, 1, 0 \rangle$ is a pseudo complemented lattice. Then the following two statements hold:*

(1) $\vdash 0 = 1$ *and* $\vdash 1 = 0$.
(2) For all x and y in L, if $x \leq y$ then $\vdash y \leq \vdash x$.

Proof. 1. Because for all x in L, $x \leq 1$ and $0 \sqcap x = 0$, it follows from Definition 3.1 that $\vdash 0 = 1$. Because for all x in L, $1 \sqcap x = 0$ if and only if $x = 0$, it follows from Definition 3.1 that $\vdash 1 = 0$.

2. Suppose x and y are arbitrary elements of L and $x \leq y$. Then,

$$x \sqcap \vdash y = (x \sqcap y) \sqcap \vdash y = x \sqcap (y \sqcap \vdash y) = x \sqcap 0 = 0.$$

Thus by Definition 3.1, $\vdash y \leq \vdash x$.   □

Theorem 3.5. *Suppose $\mathfrak{L} = \langle L, \leq, \sqcup, \sqcap, \vdash, 1, 0 \rangle$ is a boolean lattice. Define $\Rightarrow$ on L as follows: For all x and y in L ,*

$$x \Rightarrow y \ = \ \vdash x \sqcup y.$$

Then $\Rightarrow$ is the relative pseudo complementation operation on $\mathfrak{L}$ and $\vdash$ is the pseudo complementation operation on $\mathfrak{L}$.

Proof. Let a and b be arbitrary elements of L. It will be shown that $\Rightarrow$ satisfies the definition of a relative pseudo complementation operator (Definition 3.1).

Part 1. Suppose x is an arbitrary element of L such that $x \leq \vdash a \sqcup b$. Then, by distributivity,

$$x = x \sqcap (\vdash a \sqcup b) = (x \sqcap \vdash a) \sqcup (x \sqcap b).$$

Thus,

$$\begin{aligned}
a \sqcap x &= [a \sqcap (x \sqcap \vdash a)] \sqcup [a \sqcap (x \sqcap b)] \\
&= 0 \sqcup [a \sqcap x \sqcap b] \\
&\leq b.
\end{aligned}$$

Part 2. Suppose that x is an arbitrary element of L such that

$$a \sqcap x \leq b.$$

Then

$$a \sqcap x \leq a \sqcap b.$$

Because

$$x \sqcap \vdash a \leq \vdash a, \tag{3.5}$$

it follows that

$$(x \sqcap \vdash a) \sqcup (a \sqcap b) \leq \vdash a \sqcup (a \sqcap b).$$

Thus, by Equation 3.5,

$$x = x \sqcap 1 = x \sqcap (\llcorner a \sqcup a) = (x \sqcap \llcorner a) \sqcup (a \sqcap x) \leq \llcorner a \sqcup (a \sqcap b) \leq \llcorner a \sqcup b \,,$$

that is,

$$x \leq \llcorner a \sqcup b \,.$$

Parts 1 and 2 show that $\llcorner a \sqcup b$ the pseudo relative complementation operator for $\mathfrak{L}$.

Because

$$(a \Rightarrow 0) = \llcorner a \sqcup 0 = \llcorner a$$

it follows by Theorem 3.3 that $\llcorner$ is the pseudo complementation operator of $\mathfrak{L}$.   □

3.2   Pseudo Complemented Distributive Algebras of Events

For dealing with probability functions, it is convenient to have the probability functions be defined on a lattice whose domain is a set of events.

Definition 3.4 (lattice algebra of events). A lattice of the form $\mathfrak{X} = \langle \mathcal{X}, \Cup, \Cap, X, \varnothing \rangle$, where X is a nonempty set and $\mathcal{X}$ is a set of subsets of X, is called a *lattice algebra of events*.   □

Topologies are one of the most studied and important event lattices in mathematics.

Definition 3.5 (topology). Formally, a *topology* $\mathcal{T}$ is a collection of a subsets of a nonempty set X such that

- $X \in \mathcal{T}$ and $\varnothing \in \mathcal{T}$,
- $A \cap B$ is in $\mathcal{T}$ for A and B in $\mathcal{T}$,
- and for all nonempty $\mathcal{Y}$ such that $\mathcal{Y} \subseteq \mathcal{T}$,

$$\bigcup \mathcal{Y} \text{ is in } \mathcal{T} \,. \tag{3.6}$$

The set X is called the *universe of* $\mathcal{T}$. (Note that it follows from Equation 3.6 that for each A and B in $\mathcal{T}$, $A \cup B$ is in $\mathcal{T}$.)   □

As an example of a topology, let $\mathcal{E}$ be the euclidean plane. Let $\mathbf{C}$ be the set consisting of $\varnothing$ and all interiors of circles of $\mathcal{E}$, and $\mathcal{T}$ be the set consisting of $\varnothing$ and all $\bigcup \mathbf{F}$ such that $\mathbf{F} \subseteq \mathbf{C}$. Then $\mathcal{T}$ is a topology with universe $\mathcal{E}$. $\mathcal{T}$ is called the *euclidean plane topology*.

Let $\mathcal{T}$ be a topology with universe X. It follows from Equation 3.6 that $A \cup B$ is in $\mathcal{T}$ for each A and B in $\mathcal{T}$. Thus it follows from Definition 3.5 that $\langle \mathcal{T}, \subseteq, \cup, \cap, X, \varnothing \rangle$ is a lattice. The following definition describes some important topological concepts involving subsets of X.

Definition 3.6 (open, closed, boundary, closure, interior). Let $\mathcal{T}$ be a topology with universe X and A be an arbitrary subset of X. By definition,

- A is *open (in the topology $\mathcal{T}$)* if and only if $A \in \mathcal{T}$,
- and $X - A$ is *closed (in the topology $\mathcal{T}$)* if and only if $A \in \mathcal{T}$.

It follows from the above definitions that X and $\varnothing$ are open and closed. Some topologies with universe X have X and $\varnothing$ as the only elements that are both open and closed, while others have additional open and closed elements, and some topologies have all of its elements open and closed.

By definition,

- the *closure* of A, $\mathsf{cl}(A)$, is, the smallest closed set C such that $A \subseteq C$, that is,

$$\mathsf{cl}(A) = \bigcap \{B | B \text{ is closed and } A \subseteq B\};$$

- the *interior* of A, $\mathsf{int}(A)$, is the largest open set D such that $D \subseteq A$, that is,

$$\mathsf{int}(A) = \bigcup \{E | E \text{ is open and } E \subseteq A\};$$

- and the *boundary* of A, $\mathsf{bd}(A)$, is $\mathsf{cl}(A) - \mathsf{int}(A)$.

The existence of the closure, interior, and boundary of A and the property that $\mathsf{cl}[\mathsf{cl}(A)] = \mathsf{cl}(A)$ easily follow from the above definitions and the definition of "topology." □

As an example, consider the euclidean topology $\mathcal{T}$ on the euclidean plane $\mathcal{E}$. Let C be a circle in the euclidean plane, and let I be the set of points inside C. Then it follows from the definition of euclidean topology that I is an open set. The set-theoretic complement of I, $-I = \mathcal{E} - I$, is then, by Definition 3.6, a closed set. The circle C is a subset of $-I$. It is not difficult to show that

$$\mathsf{int}(-I) = (-I) - C, \quad \mathsf{cl}(I) = I \cup C, \quad \text{and} \quad \mathsf{bd}(I) = \mathsf{bd}(-I) = C.$$

Suppose c is the center of C and d is a point on C. Let $A = I - \{c\}$. Then it easily follows from the above definitions that A is open and c and

d are on the boundary of A. However, as boundary points of A, c and d are different in the following manner: $D \not\subseteq I$ for each open set D in $\mathcal{T}$ such that $d \in D$, but there exists an open set G in $\mathcal{T}$ such that $c \in G$ and $G \subseteq I$, for example, $G = I - \{c\}$. This difference is used in Chapter 5 to model an important aspect of human subjective probability behavior.

Theorem 3.6. *Suppose X is a nonempty set and $\mathcal{T}$ = the set of all subsets of X. Then $\mathcal{T}$ is a topology and each element of $\mathcal{T}$ is both open and closed.*

    Proof. Let A be an arbitrary element of $\mathcal{T}$. Then $-A$ is closed. However, because $-A \in \mathcal{T}$, $-A$ is also open.   □

Theorem 3.7. *Suppose $\mathcal{T}$ is a topology, $\mathfrak{X} = \langle \mathcal{X}, \subseteq, \cup, \cap, X, \varnothing \rangle$, $\mathcal{X} \subseteq \mathcal{T}$, and with respect to $\mathcal{T}$,*

$$\ulcorner A = \mathsf{int}(-A) \,,$$

for all A in $\mathcal{X}$. Then

$$\mathfrak{X} = \langle \mathcal{X}, \subseteq, \cup, \cap, \ulcorner, X, \varnothing \rangle$$

is a heyting lattice (Definition 3.3).

    Proof It easily follows from the definition of "relative pseudo complementation" (Definition 3.3) and properties of topologies Definition 3.5 that topological algebras of events are heyting lattices.   □

Definition 3.7 (algebras of events). The following three definitions hold:

(1) A structure $\mathfrak{X}$ is said to be a *distributive algebra of events* if and only if $\mathfrak{X}$ is a lattice of the form

$$\mathfrak{X} = \langle \mathcal{X}, \subseteq, \cup, \cap, X, \varnothing \rangle \,,$$

    where X is a nonempty set, $\mathcal{X}$ is a set of subsets of X, X and $\varnothing$ are in $\mathcal{X}$, and for all A and B in $\mathcal{X}$, $A \cup B$ and $A \cap B$ are in $\mathcal{X}$. Elements of $\mathcal{X}$ will often be called events. Note that a distributive algebra of events is a special case of a lattice algebra of events. Also note that a distributive algebra of events is distributive.

(2) $\mathfrak{X}$ is said to be a *pseudo complemented lattice algebra of events* if and only if it is a distributive algebra of events that is pseudo complemented.

(3) $\mathfrak{X} = \langle \mathcal{X}, \subseteq, \cup, \cap, X, \varnothing \rangle$ is said to be a *topological algebra of events* if and only if it is a distributive algebra of events and $\mathcal{X}$ is a topology. Note that a topological algebra of events is pseudo complemented, with the pseudo complementation of A, $\ulcorner A$, being the interior of $X - A$.   □

3.3   Lattice Representation Theorems

Theorem 2.16 of Chapter 2 showed that each lattice was isomorphic to a lattice algebra of events. Theorems like this are useful for probability theories, because they allow for event interpretations of probabilistic concepts associated with lattices. For boolean and distributive lattices, related theorems were shown by the mathematician Stone in the 1930s. Today these theorems—Theorems 2.17 of Chapter 2 and 3.8 below—are called "the Stone representation theorems". They establish isomorphisms between boolean lattices and boolean algebras of events (Theorem 2.17) and between distributive lattices and distributive algebra of events (Theorem 3.8). Stone showed the boolean version in 1936 and the distributive version in 1937. The proof for the distributive version is complicated and is presented in a separate section.

Theorem 3.8 (Stone's Representation Theorem for Distributive Lattices). *Let $\mathfrak{L} = \langle L, \leq, \sqcup, \sqcap, 1, 0 \rangle$ be a lattice. Then the following two statements are equivalent:*

(1) $\mathfrak{L}$ is distributive.
(2) $\mathfrak{L}$ is isomorphic to a lattice algebra of events.

Proof. Theorem 3.17.   □

In order to represent isomorphically pseudo complemented distributive lattices that are not boolean, a different operation on the representing lattice algebra of events is defined in terms of topological concepts in order to represent the operation of pseudo complementation.

The following theorem extends Theorem 3.8 to pseudo complemented lattices by also representing the operation of pseudo complementation.

Theorem 3.9. *Each pseudo complemented distributive lattice*

$$\mathfrak{L} = \langle L, \subseteq, \sqcup, \sqcap, \ulcorner_{\mathfrak{L}}, 1, 0 \rangle$$

is isomorphic to a pseudo complemented distributive lattice algebra of events (Definition 3.7) of the form

$$\mathfrak{X} = \langle \mathcal{X}, \subseteq, \cup, \cap, \ulcorner, X, \varnothing \rangle,$$

where for some topology $\mathcal{T}$, $\mathcal{X} \subseteq \mathcal{T}$ and, with respect to $\mathcal{T}$, for all A in $\mathcal{X}$, $\ulcorner A = int(-A)$.

Proof. Let $\mathfrak{L} = \langle L, \subseteq, \sqcup, \sqcap, \neg_{\mathfrak{L}}, 1, 0 \rangle$ be an arbitrary pseudo complemented lattice. By Theorem 3.8, let φ be an isomorphism of the distributive lattice $\mathfrak{L}' = \langle L, \subseteq, \sqcup, \sqcap, 1, 0 \rangle$ onto the lattice algebra of events, $\langle \mathcal{X}, \subseteq, \cup, \cap, X, \varnothing \rangle$. Note that the lattice $\mathfrak{L}$ is a lattice with a negative operator, but $\mathfrak{L}'$ is a lattice without a negative operator. Define $\neg$ on $\mathcal{X}$ as follows: For all A in $\mathcal{X}$,

$$\neg A = \varphi^{-1}(\neg_{\mathfrak{L}} A).$$

Then it easily follows that

$$\mathfrak{X} = \langle \mathcal{X}, \subseteq, \cup, \cap, \neg, X, \varnothing \rangle$$

is a lattice algebra of events that is isomorphic to $\mathfrak{L}$. The theorem follows by finding a topology $\mathcal{T}$ such that $\mathcal{X} \subseteq \mathcal{T}$ and showing that the operator $\neg'$ defined on $\mathcal{X}$ by,

$$\text{for all } A \text{ in } \mathcal{X}, \ \neg' A = \text{int}(-A),$$

is the pseudo complementation operator $\neg$ of $\mathfrak{X}$.

Let

$$\mathcal{T} = \{\bigcup \mathcal{B} \mid \mathcal{B} \subseteq \mathcal{X}\}.$$

Then for each A in $\mathcal{X}$, $\bigcup \{A\} = A$ is in $\mathcal{T}$. Let $\varnothing \subseteq \mathcal{D} \subseteq \mathcal{T}$. Then for each D in $\mathcal{D}$, $D \in \mathcal{X}$, and thus by the definition of $\mathcal{T}$, $\bigcup \mathcal{D}$ is in $\mathcal{T}$. Because for subsets $\mathcal{B}$ and $\mathcal{C}$ of $\mathcal{X}$,

$$\bigcup \mathcal{B} \cap \bigcup \mathcal{C} = \bigcup \{B \cap C \mid B \in \mathcal{B} \text{ and } C \in \mathcal{C}\},$$

it then easily follows from the above considerations that $\mathcal{T}$ is a topology and $\mathcal{X} \subseteq \mathcal{T}$.

Let E be an arbitrary element of $\mathcal{X}$ and, with respect to $\mathcal{T}$ and

$$\neg' E = \text{int}(-E).$$

E and $\neg E$ are in $\mathcal{T}$, because they are in $\mathcal{X}$. Thus E and $\neg E$ are open with respect to $\mathcal{T}$. Then by the definition of $\neg' E$, $E \cap \neg' E = \varnothing$. $\neg E \subseteq \neg' E$, because $\neg E \subseteq -E$ and $\neg' E$ is the interior of $-E$ and $\neg E$ is open. Thus to show the theorem, it needs to only show that $\neg' E \subseteq \neg E$. This is done by contradiction:

Suppose $x \in \neg' E$ and $x \notin \neg E$. Because $\neg' E$ is open, let $\mathcal{F} \subseteq \mathcal{X}$ be such that

$$\neg' E = \bigcup \mathcal{F}.$$

Then because x is in $\neg' E$, let F in $\mathcal{F}$ be such $x \in F$. Thus $F \in \mathcal{X}$ and $F \subseteq \neg' E$. $F \cap E = \varnothing$, because $\neg' E \subseteq -E$. Therefore, because $\neg$ is the operation pseudo complementation for the lattice $\mathfrak{X}$, $F \subseteq \neg E$, and thus x is in $\neg E$. This is a contradiction. $\square$

Definition 3.8 (topological lattice). $\mathfrak{L} = \langle L, \sqcup, \sqcap, \ulcorner, 1, 0 \rangle$ is said to be a *topological lattice* if and only if it is isomorphic to a topological algebra of events. $\square$

Theorem 3.10. *Each topological lattice is a heyting lattice.*

    Proof. By Theorem 3.7, topological algebras of events are heyting lattices. Thus, by isomorphism and the definition of topological lattice (Definition 3.8), it follows that each topological lattice of events is a heyting lattice. $\square$

    By convention, the following notations are often used in describing topological algebra of events $\mathfrak{X}$:

(*i*) $\mathfrak{X} = \langle \mathcal{X}, \subseteq, \cup, \cap, \ulcorner, X, \varnothing \rangle$, and
(*ii*) $\mathfrak{X} = \langle \mathcal{X}, \subseteq, \cup, \cap, \Rightarrow, \ulcorner, X, \varnothing \rangle$.

In (*i*) and (*ii*), $\ulcorner$ is the operator of pseudo complementation on $\mathcal{X}$, and in (*ii*), $\Rightarrow$ is the operator of relative pseudo complementation on $\mathcal{X}$. Note that for each A in $\mathcal{X}$, $\ulcorner A$ is the interior of $-A$. $\square$

    The following theorem summarizes the relationships between boolean, heyting, and pseudo complement lattices.

Theorem 3.11. *The following statements hold:*

(1) *If $\mathfrak{L}$ is a boolean lattice, then it is isomorphic to a boolean algebra of events and is a heyting lattice.*

(2) *If $\mathfrak{L}$ is a topological lattice, then it is isomorphic to a topological algebra of events and is a heyting lattice.*

(3) *If $\mathfrak{L}$ is a heyting lattice, then it is a pseudo complemented distributive lattice.*

(4) *If $\mathfrak{L}$ is a pseudo complemented distributive lattice, then it is isomorphic to a pseudo complemented distributive algebra of events.*

(5) *There exist $\mathfrak{X}_b$, $\mathfrak{X}_t$, $\mathfrak{X}_h$, and $\mathfrak{X}_r$ such that*

    (i) *$\mathfrak{X}_b$ is a boolean algebra of events but not a topological algebra of events,*

    (ii) *$\mathfrak{X}_t$ is a topological algebra of events that is not a boolean algebra of events*

    (iii) *$\mathfrak{X}_h$ is a heyting algebra of events that is not a topological algebra of events, and*

    (iv) *$\mathfrak{X}_p$ is a pseudo complemented distributive algebra of events that is not a heyting algebra of events.*

Proof. 1. Statement 1 immediately follows from Theorem 3.5.

2. Statement 2 immediate follows from Theorems 3.10 and Definition 3.8.

3. Statement 3 immediately follows from Theorem 3.3 and the definition of "pseudo complement" (Definition 3.1).

4. Statement 4 immediately follows from Theorem 3.9

$5(i)$. Let $X = \{1, 2, 3, 4, \ldots\}$, $\mathcal{F} = \{A \mid A \subseteq X$ and A is finite$\}$, and

$$\mathcal{X} = \mathcal{F} \cup \{X - A \mid A \in \mathcal{F}\}.$$

Then $\mathfrak{X}_b = \langle \mathcal{X}, \cup, \cap, -, X, \varnothing \rangle$ is a boolean algebra of events that is not a topology, because $\{2\} \in \mathcal{X}$, $\{4\} \in \mathcal{X}$, $\{6\} \in \mathcal{X}$, $\{8\} \in \mathcal{X} \ldots$, but

$$\{2, 4, 6, 8, \ldots\} \notin \mathcal{X}.$$

$5(ii)$. Let $\mathfrak{X}_t$ be any topological algebra of events such that some open set is not a closed set, for example the usual topology of open subsets of the euclidean plane. Then $\mathfrak{X}_t$ is not a boolean algebra, because it is not closed under set-theoretic complementation.

$5(iii)$. Let $\mathfrak{X}_h = \mathfrak{X}_b$, where $\mathfrak{X}_b$ is as in the proof of Statement $5(i)$. Then by Statement 1, $\mathfrak{X}_h$ is a heyting lattice algebra of events, and by Statement 5(i), $\mathfrak{X}_h$ is not a topological lattice algebra of events.

$5(iv)$. Example 3.1 provides a pseudo complemented distributive lattice algebra of events that is not a heyting lattice algebra of events. $\quad\square$

The definition of "relative pseudo complementation" is somewhat opaque about what $\Rightarrow$ has to do with "implication". However, when interpreted in terms of a topological lattice algebra of events, its relationship to classical implication $A \to B = (-A \cup B)$ of a boolean algebra of events becomes apparent.

Theorem 3.12. *Let* $\mathfrak{X} = \langle \mathcal{X}, \cup, \cap, -, \Rightarrow, X, \varnothing \rangle$ *be a topological lattice of events and A and B be elements of $\mathcal{X}$. Then $A \Rightarrow B$ is the $\subseteq$-largest element of $\mathcal{X}$ such that $(A \Rightarrow B) \subseteq (X - A) \cup B$.*

Proof. Because $-A \cap (A \Rightarrow B) \subseteq -A$ and (by Definition 3.3)

$$A \cap (A \Rightarrow B) \subseteq B,$$

it follows that

$$
\begin{aligned}
A \Rightarrow B &= (-A \cup A) \cap (A \Rightarrow B) \\
&= [-A \cap (A \Rightarrow B)] \cup [A \cap (A \Rightarrow B)] \\
&\subseteq (-A) \cup B. \quad \square
\end{aligned}
$$

Theorem 3.12 says that in a topological lattice algebra of events $\mathfrak{X} = \langle \mathcal{X}, \cup, \cap, \llcorner, \Rightarrow, X, \varnothing \rangle$,

$$A \Rightarrow B \;=\; \text{int}(A \to B),$$

where $\to$ is the standard boolean implication operator, $(-A) \cup B$ is in the boolean algebra of events $\mathfrak{B}$ that is generated by closing $\mathcal{X}$ under the operations of $\cup$, $\cap$, and $-$ (with $-$ being the set-theoretic complementation operator with respect to the event X). Also, in terms of concepts from $\mathfrak{B}$, it is easy to see that the pseudo complementation of A in $\mathfrak{X}$, $\llcorner A$, has the following definition:

$$\llcorner A \;=\; \text{int}(-A).$$

3.4   Comparison with Boolean Lattices

Pseudo complemented distributive lattices and boolean lattices share many algebraic properties. The following theorem provides some basic properties that they have in common for distributive lattices.

Theorem 3.13. *Suppose* $\mathfrak{L} = \langle L, \leq, \sqcup, \sqcap, \llcorner, 1, 0 \rangle$ *is a pseudo complemented distributive lattice. Then the following eight statements hold for all a and b in L:*

(1) $\llcorner 1 = 0$ *and* $\llcorner 0 = 1$.
(2) If $a \sqcap b = 0$, *then* $b \leq \llcorner a$.
(3) $a \sqcap \llcorner a = 0$.
(4) If $b \leq a$, *then* $\llcorner a \leq \llcorner b$.
(5) $a \leq \llcorner\llcorner a$.
(6) $\llcorner a = \llcorner\llcorner\llcorner a$.
(7) $\llcorner(a \sqcup b) = \llcorner a \sqcap \llcorner b$.
(8) $\llcorner a \sqcup \llcorner b \leq \llcorner(a \sqcap b)$.

Proof. Statements 1 to 3 are immediate from the definition of "pseudo complementation" (Definition 3.1)

4. Statement 4 follows from Theorem 3.4.

5. By Statement 3, $\llcorner a \sqcap a = 0$. Thus by Statement 2, $a \leq \llcorner\llcorner a$.

6. By Statement 5, $a \leq \llcorner\llcorner a$. Thus by Statement 4,

$$\llcorner\llcorner\llcorner a \leq \llcorner a .$$

However, by Statement 5,

$$\llcorner a \leq \llcorner\llcorner(\llcorner a) = \llcorner\llcorner\llcorner a .$$

Therefore, $\llcorner a = \llcorner\llcorner\llcorner a$.

    7. By distributivity and $x \sqcap \llcorner x = 0$ for all x in L,

$$(a \sqcup b) \sqcap (\llcorner a \sqcap \llcorner b) = [(a \sqcup b) \sqcap \llcorner a] \sqcap \llcorner b = [(a \sqcap \llcorner a) \sqcup (b \sqcap \llcorner a)] \sqcap \llcorner b$$

$$= [0 \sqcup (b \sqcap \llcorner a)] \sqcap \llcorner b = (b \sqcap \llcorner a) \sqcap \llcorner b$$

$$= (b \sqcap \llcorner b) \sqcap \llcorner a = 0 \sqcap \llcorner a = 0.$$

Because $(a \sqcup b) \sqcap (\llcorner a \sqcap \llcorner b) = 0$, it follows from Statement 2 that

$$\llcorner a \sqcap \llcorner b \leq \llcorner (a \sqcup b). \tag{3.7}$$

Because $a \leq a \sqcup b$ and $b \leq a \sqcup b$, it follows from Statement 4 that

$$\llcorner (a \sqcup b) \leq \llcorner a \quad \text{and} \quad \llcorner (a \sqcup b) \leq \llcorner b.$$

Therefore

$$\llcorner (a \sqcup b) \leq \llcorner a \sqcap \llcorner b. \tag{3.8}$$

Equations 3.7 and 3.8 show that

$$\llcorner (a \sqcup b) = \llcorner a \sqcap \llcorner b.$$

    8. From

$$a \sqcap b \leq a \quad \text{and} \quad a \sqcap b \leq b,$$

it follows from Statement 4 that

$$\llcorner a \leq \llcorner (a \sqcap b) \quad \text{and} \quad \llcorner b \leq \llcorner (a \sqcap b),$$

and thus

$$\llcorner a \sqcup \llcorner b \leq \llcorner (a \sqcap b). \quad \square$$

In distributive lattices, complementation imposes stronger algebraic constraints than pseudo complementation. Theorem 3.14 demonstrates this for some fundamental identities involving complementation that hold in all boolean lattices but fail in some pseudo complemented distributive lattices.

Theorem 3.14. *There exists a topological lattice (and therefore a pseudo complemented distributive lattice)* $\mathfrak{X} = \langle \mathcal{X}, \subseteq, \cup, \cap, \llcorner, X, \varnothing \rangle$ *such that the following three statements are true about* $\mathfrak{X}$.

(1) For some A in $\mathcal{X}$, $A \cup \llcorner A \neq X$.
(2) For some A in $\mathcal{X}$, $\llcorner\llcorner A \neq A$.
(3) For some A and B in $\mathcal{X}$, $\llcorner (A \cap B) \neq \llcorner A \cup \llcorner B$.

    Proof. Let X be the set of real numbers $\mathbb{R}$, $\mathcal{X}$ be the set of open sets of the order topology on X (i.e., the usual topology on $\mathbb{R}$ determined by the less than or equal ordering on $\mathbb{R}$), C be the infinite open interval $(0, \infty)$, and D be the infinite open interval $(-\infty, 0)$. It is easy to verify that Statement 1 follows by letting $A = C$, Statement 2 by letting $A = C \cup D$, and Statement 3 by letting $A = C$ and $B = D$. $\quad \square$

3.5 The Refutation Lattice of $\mathfrak{L}$

The pseudo complemented elements of a pseudo complemented distributive lattice $\mathfrak{L}$ themselves form an interesting lattice called "the refutation lattice of $\mathfrak{L}$".

Definition 3.9 ($\uplus$, the refutation lattice of $\mathfrak{L}$). Let

$$\mathfrak{L} = \langle L, \leq, \sqcup, \sqcap, \llcorner, 1, 0 \rangle$$

be pseudo complemented distributive lattice. By definition,

$$L_{\llcorner} = \{ \llcorner b \mid b \in L \}.$$

$\uplus$ is defined as follows: For each a and b in $L_{\llcorner}$,

$$a \uplus b = \llcorner\llcorner (a \sqcup b).$$

Then $\mathfrak{L}_{\llcorner}$ is said to be the *refutation lattice of $\mathfrak{L}$ if and only if*

$$\mathfrak{L}_{\llcorner} = \langle L_{\llcorner}, \leq, \uplus, \sqcap, \llcorner, 1, 0 \rangle. \quad \square$$

The following theorem establishes facts about $\mathfrak{L}_{\llcorner}$ that are used to show in Theorem 3.16 below that $\mathfrak{L}_{\llcorner}$ is a boolean lattice.

Theorem 3.15. *Let $\mathfrak{L} = \langle \mathcal{L}, \leq, \sqcup, \sqcap, \llcorner, 1, 0 \rangle$ be a pseudo complemented distributive lattice and $\mathfrak{L}_{\llcorner} = \langle L_{\llcorner}, \leq, \uplus, \sqcap, \llcorner, 1, 0 \rangle$ be the refutation lattice of $\mathfrak{L}$. Let a, b, and c be arbitrary elements of $L_{\llcorner}$, and a' and b' be elements of L such that $a = \llcorner a'$ and $b = \llcorner b'$. Then the following six statements hold:*

(1) $a = \llcorner\llcorner a$.
(2) $a \sqcap b$ is in $\mathfrak{L}_{\llcorner}$.
(3) If $b \leq a$, then $\llcorner\llcorner b \leq a$.
(4) $\mathfrak{L}_{\llcorner} = \langle L_{\llcorner}, \leq, \uplus, \sqcap, 1, 0 \rangle$ is a lattice with $a \sqcap b$ as the meet of a and b, and $a \uplus b$ as the join of a and b.
(5) $(a \sqcap b) \uplus (a \sqcap c) = a \sqcap (b \uplus c)$.
(6) $a \sqcup \llcorner a = 1$.

Proof. 1. Because $a = \llcorner a'$, it follows from Statement 6 of Theorem 3.13 that

$$\llcorner\llcorner a = \llcorner\llcorner\llcorner a' = \llcorner a' = a.$$

2. By Statement 7 of Theorem 3.13,

$$a \sqcap b = \llcorner a' \sqcap \llcorner b' = \llcorner (a' \sqcup b').$$

3. By Statement 1, $b = \ulcorner\ulcorner b$.

4. Because $1 = \neg\, 0$ and $0 = \neg\, 1$, 1 and 0 are in $L_\ulcorner$. Because 1 and 0 are, respectively, the $\leq$-maximal and $\leq$-minimal elements of L, they are, respectively, the $\leq$-maximal and $\leq$-minimal elements of $L_\ulcorner$.

By Statement 2, $a \sqcap b$ is in $L_\ulcorner$. Because $a \sqcap b$ is the meet in $\mathfrak{L}$ and $L_\ulcorner \subseteq \mathcal{X}$, it follows that $a \sqcap b$ is the meet in $\mathfrak{L}_\ulcorner$.

Let d be an arbitrary element in $L_\ulcorner$ such that

$$a \leq d \text{ and } b \leq d. \tag{3.9}$$

(There exists such a d, because Equation 3.9 holds for $d = 1$.) To show that $a \uplus b$ is the join of a and b in $\mathfrak{L}_\ulcorner$, it needs only to be shown that $a \leq a \uplus b \leq d$ and $b \leq a \uplus b \leq d$. By the choice of d, $a \sqcup b \leq d$. By two applications of Statement 4 of Theorem 3.13,

$$\ulcorner\ulcorner (a \sqcup b) \leq \ulcorner\ulcorner d.$$

By Statement 5 of Theorem 3.13, $a \sqcup b \leq \ulcorner\ulcorner (a \sqcup b)$, and by Statement 1, $\ulcorner\ulcorner d = d$. Thus,

$$a \sqcup b \leq \ulcorner\ulcorner (a \sqcup b) = a \uplus b \leq d.$$

Therefore,

$$a \leq a \uplus b \leq d \text{ and } b \leq a \uplus b \leq d.$$

Thus $a \uplus b$ is the join of a and b in $\mathfrak{L}_\ulcorner$.

5. The following series of equations hold:

$$
\begin{aligned}
(a \sqcap b) \uplus (a \sqcap c) &= \ulcorner\ulcorner [(a \sqcap b) \sqcup (a \sqcap c)] \\
&= \ulcorner\ulcorner [a \sqcap (b \sqcup c)] \\
&\geq \ulcorner [\ulcorner a \sqcup \ulcorner (b \sqcup c)] \quad \text{(Statement 8 of Theorem 3.13)} \\
&= [\ulcorner\ulcorner a \sqcap \ulcorner\ulcorner (b \sqcup c)] \quad \text{(Statement 7 of Theorem 3.13)} \\
&= [\ulcorner\ulcorner a \sqcap (b \uplus c)] \\
&= a \sqcap (b \uplus c)] \quad \text{(Statement 1)},
\end{aligned}
$$

that is,

$$(a \sqcap b) \uplus (a \sqcap c) \geq a \sqcap (b \uplus c). \tag{3.10}$$

Because,

$$(a \sqcap b) \sqcup (a \sqcap c) = a \sqcap (b \sqcup c)$$
$$\leq a \sqcap \ulcorner\urcorner (b \sqcup c) \quad \text{(Statement 5 of Theorem 3.13)}$$
$$= a \sqcap (b \uplus c),$$

it follows that

$$(a \sqcap b) \sqcup (a \sqcap c) \leq a \sqcap (b \uplus c).$$

Then

$$a \sqcap b \leq a \sqcap (b \uplus c) \quad \text{and} \quad a \sqcap c \leq a \sqcap (b \uplus c).$$

Thus, because by Statement 4 $(a \sqcap b) \uplus (a \sqcap c)$ is the $\leq$-smallest element in $L_\ulcorner$ such that

$$a \sqcap b \leq (a \sqcap b) \uplus (a \sqcap c) \quad \text{and} \quad a \sqcap c \leq (a \sqcap b) \uplus (a \sqcap c),$$

it then follows that

$$(a \sqcap b) \uplus (a \sqcap c) \leq a \sqcap (b \uplus c). \tag{3.11}$$

Equations 3.10 and 3.11 show

$$(a \sqcap b) \uplus (a \sqcap c) = a \sqcap (b \uplus c).$$

6. The following series of equations hold:

$$a \sqcup \ulcorner a = \ulcorner\urcorner (a \sqcup \ulcorner a) \quad \text{(Statement 1)}$$
$$= \ulcorner (\ulcorner a \sqcap \ulcorner\urcorner a) \quad \text{(Statement 7 of Theorem 3.13)}$$
$$= \ulcorner (\ulcorner a \sqcap a) \quad \text{(Statement 1)}$$
$$= \ulcorner 0 \quad \text{(definition of ``}\ulcorner\text{'')}$$
$$= 1 \quad \text{(definition of ``}\ulcorner\text{'')}. \quad \square$$

Theorem 3.16. *Let* $\mathfrak{L} = \langle \mathcal{L}, \leq, \sqcup, \sqcap, \ulcorner, 1, 0 \rangle$ *be pseudo complemented distributive lattice. Then the refutation lattice of* $\mathfrak{L}$,

$$\mathfrak{L}_\ulcorner = \langle L_\ulcorner, \leq, \uplus, \sqcap, \ulcorner, 1, 0 \rangle,$$

is a boolean lattice.

Proof. $\leq$ partially orders $L_\ulcorner$, and 1 and 0 are respectively the maximal and minimal elements of $L_\ulcorner$ with respect to $\leq$. By Statement 4 of Theorem 3.15, $\sqcap$ and $\uplus$ are respectively the meet and join operations of $\mathfrak{L}_\ulcorner$. By Statement 5 of Theorem 3.15, $\sqcap$ distributes over $\uplus$, and by the definition of "pseudo complement" and Statement 6 of Theorem 3.15, $\ulcorner$ is a complement operator of $\mathfrak{L}_\ulcorner$. Thus $\mathfrak{L}_\ulcorner$ is a boolean lattice. $\square$

Let $\mathfrak{X} = \langle \mathcal{X}, \cup, \cap, \ulcorner, X, \varnothing \rangle$ be a topological lattice. Then by Theorem 3.16, the refutation lattice of $\mathfrak{X}$,

$$\mathfrak{X}_\ulcorner = \langle \mathcal{X}, \Cup, \cap, \ulcorner, \varnothing, \rangle \,,$$

is a boolean lattice of events. $\mathfrak{X}_\ulcorner$ is not necessarily a boolean algebra of events, because $\Cup$ is not necessarily $\cup$. The case where $\Cup = \cup$ is called a *stone lattice*.

Definition 3.10 (stone lattice of $\mathfrak{L}$). Let $\mathfrak{L} = \langle \mathcal{L}, \leq, \sqcup, \sqcap, \ulcorner, 1, 0 \rangle$ be pseudo complemented distributive lattice. Then the refutation lattice of $\mathfrak{L}$,

$$\mathfrak{L}_\ulcorner = \langle L_\ulcorner, \leq, \uplus, \sqcap, \ulcorner, 1, 0 \rangle \,,$$

is said to be a *stone lattice* if and only if $\sqcup = \uplus$. $\square$

3.6   Proof of Stone's Representation Theorem for Distributive Lattices

Stone (1937) showed that each distributive lattice is isomorphic to a distributive lattice algebra of events. The proof presented here for this result follows the presentation of Rasiowa and Sikorski (1963), which follows the presentation of Stone (1937).

Stone's method of proof is based on the algebraic concept of "filter." Filters, especially maximally large filters ("ultrafilters") play an important role in many representation theorems.

Definition 3.11. Let $\mathfrak{L} = \langle L, \leq, \sqcup, \sqcap, 1, 0 \rangle$ be a lattice.

$\mathcal{F}$ is said to be a *filter (of $\mathfrak{L}$)* if and only if $\mathcal{F}$ is a nonempty subset of L and for all a and b in L,

(1) $0 \notin \mathcal{F}$.
(2) if $a \in \mathcal{F}$ and $b \in \mathcal{F}$, then $a \sqcap b \in \mathcal{F}$, and
(3) if $a \in \mathcal{F}$ and $a \leq b$, then $b \in \mathcal{F}$.

For each $a \neq 0$ in L, $\{b \,|\, a \leq b\}$ is called the *principal filter generated by* a. It easily follows that for each $a \neq 0$ in L, the principal filter generated by a is a filter of $\mathfrak{L}$. $\square$

Definition 3.12. Let $\mathfrak{L} = \langle L, \leq, \sqcup, \sqcap, 1, 0 \rangle$ be a lattice. Then $\mathbf{C}$ is called a *chain of filters* of $\mathfrak{L}$ if and only if $\mathbf{C}$ is a nonempty set of filters such that for all $\mathcal{F}$ and $\mathcal{G}$ in $\mathbf{C}$, either $\mathcal{F} \subseteq \mathcal{G}$ or $\mathcal{G} \subseteq \mathcal{F}$. $\square$

Lemma 3.1. *Let* $\mathfrak{L} = \langle L, \leq, \sqcup, \sqcap, 1, 0 \rangle$ *be a lattice and* C *be a chain of filters of* $\mathfrak{L}$. *Then* $\bigcup \mathcal{F}$ *is a filter of* $\mathfrak{L}$.

    Proof. The proof is straightforward and is left to the reader.   □

Definition 3.13. Let $\mathfrak{L} = \langle L, \leq, \sqcup, \sqcap, 1, 0 \rangle$ be a lattice. Then $\mathcal{F}$ is said to be an *ultrafilter* of $\mathfrak{L}$ if and only if $\mathcal{F}$ is a filter of $\mathfrak{L}$ and for each filter $\mathcal{G}$ of $\mathfrak{L}$, if $\mathcal{F} \subseteq \mathcal{G}$ then $\mathcal{F} = \mathcal{G}$.   □

Lemma 3.2. *Let* $\mathfrak{L} = \langle L, \leq, \sqcup, \sqcap, 1, 0 \rangle$ *be a lattice. Then each filter of* $\mathfrak{L}$ *is contained in an ultrafilter of* $\mathfrak{L}$.

    Proof. Let $\mathcal{F}$ be a filter on $\mathfrak{L}$ and

$$\mathbf{G} = \{\mathcal{G} \,|\, \mathcal{G} \text{ is a filter of } \mathfrak{L} \text{ and } \mathcal{F} \subseteq \mathcal{G}\}.$$

Then $\mathbf{G} \neq \varnothing$, because $\mathcal{F} \in \mathbf{G}$, and if $\mathbf{C}$ is a chain of filters such that $\mathbf{C} \subseteq \mathbf{G}$, then $\bigcup \mathbf{C}$ is in $\mathbf{G}$. Thus by Zorn's Lemma, $\mathbf{G}$ has a $\subseteq$-maximal element, and this maximal element is an ultrafilter containing $\mathcal{F}$.   □

Definition 3.14. Let $\mathfrak{L} = \langle L, \leq, \sqcup, \sqcap, 1, 0 \rangle$ be a lattice. Then $\mathcal{F}$ is said to be a *prime* filter of $\mathfrak{L}$ if and only if $\mathcal{F}$ is a filter of $\mathfrak{L}$ and for each a and b in L,

$$\text{if } a \sqcup b \in \mathcal{F} \text{ then } a \in \mathcal{F} \text{ or } b \in \mathcal{F}. \quad □$$

Lemma 3.3. *Let* $\mathfrak{L} = \langle L, \leq, \sqcup, \sqcap, 1, 0 \rangle$ *be a distributive lattice. Then each ultrafilter of* $\mathfrak{L}$ *is prime.*

    Proof. Suppose $\mathcal{F}$ is an ultrafilter of $\mathfrak{L}$ that is not prime. A contradiction will be shown. Let a and b be elements of L such that

$$a \sqcup b \in \mathcal{F}, \ a \notin \mathcal{F}, \text{ and } b \notin \mathcal{F}.$$

Let

$$\mathcal{G} = \{x \,|\, \text{there exists } c \in \mathcal{F} \text{ such that } a \sqcap c \leq x\}.$$

It easily follows that if d and e are in $\mathcal{G}$, then $d \sqcap e$ is in $\mathcal{G}$, and for all y in L, if $d \leq y$, then y is in $\mathcal{G}$. Thus to show that $\mathcal{G}$ is a filter, it needs only be shown that $0 \notin \mathcal{G}$. Suppose 0 were an element of $\mathcal{G}$. A contradiction will be shown. Then by the definition of $\mathcal{G}$, let c in $\mathcal{F}$ be such that $a \sqcap c = 0$. Then, because $(a \sqcup b) \in \mathcal{F}$ and $\mathfrak{L}$ is distributive, it follows that

$$c \sqcap (a \sqcup b) = (c \sqcap a) \sqcup (c \sqcap b) = 0 \sqcup (c \sqcap b) = c \sqcap b$$

is in $\mathcal{F}$. Then, because $c \sqcap b \leq b$, b is in $\mathcal{F}$, contradicting $b \notin \mathcal{F}$. Thus $\mathcal{G}$ is a filter. It is immediate that $\mathcal{F} \subseteq \mathcal{G}$. $a \in \mathcal{G}$, because $1 \in \mathcal{G}$ and $a = a \sqcap 1$. Thus, because by hypothesis $a \notin \mathcal{F}$, it follows that $\mathcal{F} \subset \mathcal{G}$, contradicting that $\mathcal{F}$ is an ultrafilter.   □

Lemma 3.4. *Let* $\mathfrak{L} = \langle L, \leq, \sqcup, \sqcap, 1, 0 \rangle$ *be a distributive lattice and a and b be elements of L such that $b \not\leq a$. Then there exists a prime filter $\mathcal{F}$ such that*

$$a \notin \mathcal{F} \quad and \quad b \in \mathcal{F}.$$

Proof. Let $\mathbf{G}$ be the set of all filters $\mathcal{G}$ such that

$$a \notin \mathcal{G} \quad and \quad b \in \mathcal{G}.$$

Then $\mathbf{G}$ is nonempty because it contains the principal filter generated by b. Using Lemma 3.1, it immediately follows that $\bigcup \mathbf{C}$ is in $\mathbf{G}$ for each chain of filters $\mathbf{C}$ of $\mathfrak{L}$ such that $\mathbf{C} \subseteq \mathbf{G}$. Thus by Zorn's Lemma, let $\mathcal{H}$ be a largest element of $\mathbf{G}$ in terms of the $\subseteq$-ordering. Then, by the definitions of $\mathbf{G}$ and $\mathcal{H}$,

$$a \notin \mathcal{H} \quad and \quad b \in \mathcal{H}. \tag{3.12}$$

Thus to show the lemma, it needs only be shown that $\mathcal{H}$ is prime. This will be done by contradiction. Suppose $\mathcal{H}$ were not prime. Let c and d be elements of L such that

$$c \sqcup d \in \mathcal{H}, \ c \notin \mathcal{H}, \quad and \quad d \notin \mathcal{H}.$$

Let

$$\mathcal{H}_c = \{x \mid x \in L \text{ and there exists } h \text{ in } \mathcal{H} \text{ such that } c \sqcap h \leq x\}$$

and

$$\mathcal{H}_d = \{x \mid x \in L \text{ and there exists } h \text{ in } \mathcal{H} \text{ such that } d \sqcap h \leq x\}.$$

It will be shown by contradiction that either $a \notin \mathcal{H}_c$ or $a \notin \mathcal{H}_d$:

Suppose $a \in \mathcal{H}_c$ and $a \in \mathcal{H}_d$. Then by the definitions of $\mathcal{H}_c$ and $\mathcal{H}_d$, c_1 and d_1 in $\mathcal{H}$ can be found such that

$$c \sqcap c_1 \leq a \quad and \quad d \sqcap d_1 \leq a.$$

Let $e = c_1 \sqcap d_1$. Then e is in $\mathcal{H}$, and

$$c \sqcap e \leq a \quad and \quad d \sqcap e \leq a.$$

Therefore,

$$a \geq (c \sqcap e) \sqcup (d \sqcap e) = (c \sqcup d) \sqcap e \in \mathcal{H},$$

which implies $a \in \mathcal{H}$, because $c \sqcup d$ and e are in $\mathcal{H}$. This contradicts $a \notin \mathcal{H}$.

Thus, without loss of generality it can be assumed that $a \notin \mathcal{H}_c$. It is immediate from the definition of $\mathcal{H}_c$ that $\mathcal{H}_c$ satisfies all the conditions of a filter of $\mathfrak{L}$ except for, perhaps, the exclusion of 0 from being an element of

$\mathcal{H}_c$. $0 \notin \mathcal{H}_c$, because if it were, then a would be an element of $\mathcal{H}_c$ because then it would follow that $0 \leq a$, contradicting the above supposition that $a \notin \mathcal{H}_c$. By the definition of $\mathcal{H}_c$, $\mathcal{H} \subseteq \mathcal{H}_c$. By Equation 3.12, $b \in \mathcal{H}$, and thus $b \in \mathcal{H}_c$. By the choice of c, $c \notin \mathcal{H}$. However, because $c \in \mathcal{H}_c$, it then follows that $\mathcal{H} \subset \mathcal{H}_c$. This together with the facts above that establish $\mathcal{H}_c$ is in **G** contradicts the choice of $\mathcal{H}$ as a $\subseteq$-largest element of **G**. $\qquad \square$

Theorem 3.17 (Stone's Representation Theorem for Distributive Lattices).   *Let* $\mathfrak{L} = \langle L, \leq, \sqcup, \sqcap, 1, 0 \rangle$ *be a lattice. Then the following two statements are equivalent:*

(1) $\mathfrak{L}$ *is distributive.*
(2) $\mathfrak{L}$ *is isomorphic to a distributive lattice algebra of events.*

Proof. Because each distributive lattice algebra of events is distributive, it is immediate that Statement (2) implies Statement (1).

Suppose Statement (1). Let **P** be the set of prime filters of $\mathfrak{L}$. For each a in L, let
$$\varphi(a) = \{\mathcal{F} \mid \mathcal{F} \in \mathbf{P} \text{ and } a \in \mathcal{F}\},$$
and
$$\boldsymbol{\mathcal{P}} = \{\varphi(a) \mid a \in L\}.$$
It will be shown that φ is an isomorphism of $\langle L, \leq, \sqcup, \sqcap, 1, 0 \rangle$ onto
$$\langle \boldsymbol{\mathcal{P}}, \subseteq, \cup, \cap, \mathbf{P}, \varnothing \rangle.$$
Let a and b be arbitrary elements of L.

To show φ is one-to-one, suppose $a \neq b$. Then either $a \nleq b$ or $b \nleq a$. By Lemma 3.4, there is a prime filter $\mathcal{F}$ of $\mathfrak{L}$ such that $\mathcal{F}$ is in exactly one of the sets $\varphi(a)$, $\varphi(b)$. Therefore, $\varphi(a) \neq \varphi(b)$.

By the definition of $\boldsymbol{\mathcal{P}}$, φ is onto $\boldsymbol{\mathcal{P}}$.

Because 1 is an element of each filter of $\mathfrak{L}$, it follows that $\varphi(1) = \mathbf{P}$, and because 0 is not an element of any filter of $\mathfrak{L}$, $\varphi(0) = \varnothing$.

Suppose $\mathcal{F}$ is an arbitrary element of $\varphi(a \sqcup b)$. Then $\mathcal{F}$ is a prime filter of $\mathfrak{L}$ and $(a \sqcup b) \in \mathcal{F}$. Because $\mathcal{F}$ is prime, then either $a \in \mathcal{F}$ or $b \in \mathcal{F}$. Thus either $\mathcal{F} \in \varphi(a)$ or $\mathcal{F} \in \varphi(b)$, that is, $\mathcal{F} \in [\varphi(a) \cup \varphi(b)]$. This argument shows
$$\varphi(a \sqcup b) \subseteq \varphi(a) \cup \varphi(b). \tag{3.13}$$
Suppose $\mathcal{G}$ is an arbitrary element of $\varphi(a) \cup \varphi(b)$. Then either $\mathcal{G} \in \varphi(a)$ or $\mathcal{G} \in \varphi(b)$, that is, either $a \in \mathcal{G}$ or $b \in \mathcal{G}$. Thus, because $\mathcal{G}$ is a filter, $(a \sqcup b) \in \mathcal{G}$, and therefore, $\mathcal{G} \in \varphi(a \sqcup b)$. This argument shows
$$\varphi(a) \cup \varphi(b) \subseteq \varphi(a \sqcup b). \tag{3.14}$$

Thus, by Equations 3.13 and 3.14, $\varphi(a \sqcup b) = \varphi(a) \cup \varphi(b)$.

Suppose $\mathcal{F}$ is an arbitrary element of $\varphi(a \sqcap b)$. Then $\mathcal{F}$ is a prime filter of $\mathfrak{L}$ and $(a \sqcap b) \in \mathcal{F}$. Then, because $\mathcal{F}$ is a filter of $\mathfrak{L}$ and $a \sqcap b \leq a$ and $a \sqcap b \leq b$, it follows that $a \in \mathcal{F}$ and $b \in \mathcal{F}$, that is, $\mathcal{F} \in \varphi(a)$ and $\mathcal{F} \in \varphi(b)$, and thus $\mathcal{F} \in \varphi(a) \cap \varphi(b)$. This argument shows

$$\varphi(a \sqcap b) \subseteq \varphi(a) \cap \varphi(b). \tag{3.15}$$

Suppose $\mathcal{G}$ is an arbitrary element of $\varphi(a) \cap \varphi(b)$. Then $\mathcal{G} \in \varphi(a)$ and $\mathcal{G} \in \varphi(b)$; that is, $a \in \mathcal{G}$ and $b \in \mathcal{G}$. Thus, because $\mathcal{G}$ is a filter, $a \sqcap b \in \mathcal{G}$, and therefore $\mathcal{G} \in \varphi(a \sqcap b)$. This argument shows

$$\varphi(a) \cap \varphi(b) \subseteq \varphi(a \sqcap b). \tag{3.16}$$

Thus by Equations 3.15 and 3.16, $\varphi(a \sqcap b) = \varphi(a) \cap \varphi(b)$. $\quad\square$

Chapter 4

Probability and Coherence

Traditional probability theory is the standard method for dealing with uncertainty in science and philosophy. It has been widely and successfully applied in a variety of domains. Nevertheless, there are important areas where it appears to be too restrictive. This and later chapters use some of the concepts and tools of previous chapters to generalize traditional probability theory and apply the generalizations to important probabilistic situations.

The first generalization involves the concept of "probability function". Traditional probability theory has probability functions that take values in the real interval [0,1]. This is too restrictive to appropriately model decision situations where there is clearly a better choice to an alternative but no way to express this using real numbers, because the better choice is only "infinitesimally better". In probability theory a similar issue arises where one wants to distinguish the uncertainty in A and B, where $A \subset B$ but A and B have the same traditional probability. A natural way to deal with the latter is to extend the system of real numbers to include infinitesimal elements so that the probability of B becomes infinitesimally greater than the probability A. This chapter presents a foundation for such an approach.

The second generalization involves the Dutch Book Argument. Science and philosophy employs the Dutch Book Argument to show that traditional probability theory is *the* rational theory for the numerical representation of uncertainty. This chapter generalizes the Argument in four ways:

(1) It shows that the Argument's assumption of an underlying boolean algebra of events can be replaced by a distributive algebra of events.
(2) It views the Argument's concept of "rationality" as too restrictive and generalizes it.
(3) It replaces the Argument's assumed quantitative model with a qualitative preference model.

(4) It fills a conceptual gap in the usual presentations of the Argument involving whether or not rationality would be violated if additional events were included in the underlying event algebra.

The third generalization presents a qualitative measurement-theoretic foundation for probability theory such that

- its probability function is finitely additive and can have values that differ by a positive infinitesimal;
- its event space is distributive (including the possibility that it is boolean); and
- its key qualitative concept bears a close relationship to the key concept used in the Dutch Book Argument.

4.1   L-Probabilistic Lattices

L-Probabilistic lattices generalize traditional probability theory to a broader class of probability functions. The generalization has the two following characteristics:

- The "additivity" of disjoint events is generalized so that it applies to joins of events that are not necessarily joins of disjoint events.
- Various non-null events can take on positive infinitesimal values.

This generalization allows for a cleaner algebraic development of probabilistic concepts and the production of sharper theorems. The main result about L-probabilistic lattices presented in this section is that they are necessarily modular, and they are boolean if and only if they are uniquely complemented.

Lattices are abstract algebraic structures with many substantive empirical and mathematical interpretations. The two mathematical interpretations of "lattice" used throughout this book are (i) lattices as event spaces and (ii) lattices as propositional logics. In the event space interpretation, the lattice is taken as a set lattice of the form $\mathfrak{X} = \langle \mathcal{X}, \subseteq, \mathbb{U}, \mathbb{n}, X, \varnothing \rangle$. (Some of the event space interpretations presented in this book employ a more general meet operation $\mathbb{n}$ than is required in the set representation theorem given in Theorem 2.16 that uses $\cap$ as its meet operation.) The elements of $\mathcal{X}$ are usually called *events*, X the *sure event* or the *set of states of the world*, and elements of X are called *states of the world* or *possible states of the world*. One element of X is taken as the *actual state of the world*. An event A in $\mathcal{X}$ is said *to occur* if and only if the actual state of the world is in A. Probability functions measure the likeliness of the occurrence of events in $\mathcal{X}$.

A lattice $\mathfrak{L} = \langle L, \sqcup, \sqcap, \vdash, 1, 0 \rangle$ with a negative operator can be viewed as a propositional logic if an implication operator $\rightarrow$ is appropriately defined on L. For example, if $\mathfrak{L}$ is a classical logic, then $\mathfrak{L}$ is boolean and $(a \rightarrow b)$ is defined as $(\vdash a) \sqcup b$. If $\mathfrak{L}$ is a propositional logic and a is in L, then a probability function on $\mathfrak{L}$ is interpreted as a measure of the degree belief that a is true.

4.1.1 *Infinitesimals*

Consider the following situation: $\mathbb{P}$ is the traditional probability function on $[0, 1]$ determined by a uniform distribution. Then

$$\mathbb{P}(\{1/2\}) = 0 \ \text{ and } \ \mathbb{P}(\{1/2\} \cup \{1/3\}) = 0 \,.$$

However, even though $\mathbb{P}(\{1/2\}) = 0$, the event $\{1/2\}$ has some chance of occurring—an infinitesimal chance—and the event $\{1/2\} \cup \{1/3\}$ has twice that infinitesimal chance of occurring.

The concept of a set having a "probability zero" does not distinguish between the possibility of the event $\{1/2\}$ occurring and the impossibility of the null event $\varnothing$ of occurring. While in most applications of probability theory such a distinction is not needed, in some theoretical uses of probability it is, especially in connecting probability theory to logic. Also, as a practical matter for showing theorems in lattice theory, extending the concept of probability so that probabilities can have infinitesimals as values provides for additional methods of proof that are novel and powerful.

Probability functions having infinitesimal values is not a modern concept. Until the later part of the 19th century, infinitesimals were in wide use in mathematics. They fell out of favor because at the time they lack a rigorous foundation. However, they were reintroduced into modern mathematics in Robinson (1966), which provided a rigorous foundation for various classical uses infinitesimal and infinite quantities. This chapter employs only very simple, algebraic properties involving the addition, multiplication, and ordering of infinitesimals, which are presented axiomatically.

Many of the properties of addition, multiplication, and the ordering of real and rational numbers are summarized by the concept of a "totally ordered field."

Definition 4.1 (totally ordered field). $\mathfrak{F} = \langle F, \leq, +, \cdot, 1, 0 \rangle$ is said to be a *totally ordered field* if and only if $\leq$ is a binary relation on F, $+$ and $\cdot$ are binary operations on F, 1 and 0 are elements of F, and the following four statements hold:

(1) $\leq$ is a total odering on F.

(2) $\langle F, \leq, 0 \rangle$ satisfies the following five properties for all x, y, and z in F:

  - $x + (y + z) = (x + y) + z$;
  - $x + y = y + x$;
  - $x + 0 = x$;
  - there exists u (denoted by $-x$) such that $x + u = 0$;
  - and $x \leq y$ iff $x + z \leq y + z$.

(3) Let $F^+ = \{x \mid 0 < x\}$. Then $\langle F^+, \cdot, 1 \rangle$ satisfies the following five properties for all x, y, and z in F^+:

  - $x \cdot y$ is in F^+;
  - $x \cdot (y \cdot z) = (x \cdot y) \cdot z$;
  - $x \cdot y = y \cdot x$;
  - $x \cdot 1 = x$; and
  - $x \leq y$ iff $x \cdot z \leq y \cdot z$.

(4) $\langle F, \leq, +, \cdot, 1, 0 \rangle$ satisfies the following four properties for all x, y, and z in F:

  - $1 \neq 0$;
  - $x \cdot 0 = 0$;
  - if $x \neq 0$, then there exists v (denoted by x^{-1}) such that $x \cdot v = 1$;
  - and $x \cdot (y + z) = (x \cdot y) + (x \cdot z)$. $\square$

$\mathfrak{R} = \langle \mathbb{R}, \leq, +, 1, 0 \rangle$, where $\mathbb{R}$ is the set of real numbers, is a totally ordered field. There are many more examples of totally ordered fields. What distinguishes the totally ordered field of real numbers $\mathfrak{R}$ from other totally ordered fields is that $\mathfrak{R}$ is *Dedekind complete,* (Definition 2.6) that is, the totally ordered set $\langle \mathbb{R}, \leq \rangle$ is Dedekind complete:

Theorem 4.1. *Each Dedekind complete totally ordered field is isomorphic to the totally ordered field of the reals,* $\langle \mathbb{R}, \leq, +, \cdot, 1, 0 \rangle$.

  Proof. This is a well-known theorem of algebra. $\square$

Definition 4.2 (totally ordered field extension of the reals). $^{\star}\mathfrak{R} = \langle {^{\star}\mathbb{R}}, \leq, +, 1, 0 \rangle$ is said to be an *ordered field extension of the reals* if and only if

- $^\star\mathfrak{R}$ is a totally ordered field,
- $\mathbb{R} \subseteq {}^\star\mathbb{R}$, 1 (as an element of $^\star\mathbb{R}$) is in $\mathbb{R}$, and 0 (as an element of $^\star\mathbb{R}$) is in $\mathbb{R}$,
- $+$ (as an operation on $^\star\mathbb{R}$), when restricted to $\mathbb{R}$, is the operation of addition on $\mathbb{R}$,
- and $\cdot$ (as an operation on $^\star\mathbb{R}$), when restricted to $\mathbb{R}$, is the operation of multiplication on $\mathbb{R}$.

Suppose $^\star\mathfrak{R} = \langle {}^\star\mathbb{R}, \leq, +, \cdot, 1, 0 \rangle$ is an ordered field extension of the reals. Then $^\star\mathfrak{R}$ is said to be *proper* if and only if $\mathfrak{R} \subset {}^\star\mathfrak{R}$.　□

Note by Definition 4.2, the totally ordered field of real numbers is an ordered field extension of itself. It is, of course, a non-proper extension.

The following theorem is a well-known in mathematics.

Theorem 4.2. *There exists a proper totally ordered field extension of the reals.*[1]　□

Definition 4.3 (infinitesimal and finite elements). Let

$$^\star\mathfrak{R} = \langle {}^\star\mathbb{R}, \leq, +, 1, 0 \rangle$$

be an ordered field extension of the reals.

For each α in $^\star\mathfrak{R}$, $|\alpha|$ is defined as α if $0 \leq \alpha$, and as $-\alpha$ if $0 \leq -\alpha$.

An element β of $^\star\mathfrak{R}$ is said to be *infinitesimal* if and only if for each r in $\mathbb{R}^+$, $|\beta| < r$. An element γ of $^\star\mathfrak{R}$ is said to be *finite* if and only if for some r and s in $\mathbb{R}$, $r \leq \beta \leq s$. Obviously, 0 is infinitesimal, each infinitesimal element is finite, and each element of $\mathbb{R}$ is finite.

$^\star\mathbb{R}$ is called the *extended real numbers* or just *extended reals* for short, $^\star\mathbb{R}^+$ the *extended positive reals*, and $^\star[0,1]$ the *extended closed unit interval*, where

$$^\star[0,1] = \{x \mid x \in {}^\star\mathfrak{R} \text{ and } 0 \leq x \leq 1\}.$$

---

[1] Theorem 4.2 has the following simple proof by the compactness of mathematical logic: One formulates a first-order language for the totally ordered field of real numbers, $\mathfrak{R}$, with a constant symbol $\mathbf{r}$ for each real number r. Let $\mathcal{T}$ be the set of true sentences in this language about $\mathfrak{R}$. One adds a new constant symbol $\mathbf{c}$ to the set of symbols and sentences of the form $\mathbf{c} \neq \mathbf{r}$, for each real number r, to $\mathcal{T}$ to form a larger set of sentences $\mathcal{T}'$. Because the set of real numbers is infinite, it is easy to show that each finite subset of $\mathcal{T}'$ is consistent and thus, by the compactness theorem of mathematical logic, has a model $^\star\mathfrak{R}$. Because $\mathfrak{R}$ being a totally ordered field is a first-order consequence of $\mathcal{T}$, it follows that $^\star\mathfrak{R}$ is a totally ordered field. It follows from $\mathcal{T}$ that $^\star\mathfrak{R}$ contains a submodel that is isomorphic to $\mathfrak{R}$, which, without loss of generality, we may assume to be $\mathfrak{R}$. Because $\mathbf{c} \neq \mathbf{r}$ for each r in $\mathbb{R}$, the interpretation of $\mathbf{c}$ in $^\star\mathfrak{R}$ is not in $\mathbb{R}$, that is $^\star\mathfrak{R}$ is a proper extension of $\mathfrak{R}$.

Note that the definition of "extended real numbers" includes the case where $^\star\mathbb{R} = \mathbb{R}$. □

Definition 4.4 (standard part). Let $\langle {}^\star\mathbb{R}, \leq, +, \cdot, 0, 1 \rangle$ be an ordered field extension of the reals and α be a finite element of $^\star\mathbb{R}$. Let

$$A = \{t \mid t \in \mathbb{R} \text{ and } t < \alpha\} \text{ and } B = \{t \mid t \in \mathbb{R} \text{ and } \alpha \leq t\}.$$

Then, because α is finite, (A, B) is a Dedekind cut of $\langle \mathbb{R}, \leq \rangle$. Let s be the cut element of (A, B). Then, s is in $\mathfrak{R}$, and, by definition, $^\circ\alpha = s$. $^\circ\alpha$ is called the *standard part* of α. □

Theorem 4.3. *Let $\langle {}^\star\mathbb{R}, \leq, +, \cdot, 0, 1 \rangle$ be an ordered field extension of the reals and α be a finite element of $^\star\mathbb{R}$. Then $|{}^\circ\alpha - \alpha|$ is infinitesimal.*
  Proof. Let r be an arbitrary element of $\mathbb{R}^+$ and

$$A = \{x \mid x \in \mathbb{R}^+ \text{ and } x \leq \alpha\} \text{ and } B = \{x \mid x \in \mathbb{R}^+ \text{ and } x > \alpha\}.$$

Because $^\circ\alpha$ is the cut element of the Dedekind cut (A, B),

$$(^\circ\alpha - r) \in A \quad \text{and} \quad (^\circ\alpha + r) \in B.$$

But then,

$$^\circ\alpha - r \leq \alpha \leq {}^\circ\alpha + r.$$

In other words,

$$|{}^\circ\alpha - \alpha| \leq r.$$

Because r is an arbitrary element of $\mathbb{R}^+$, it then follows from Definition 4.3 that $|{}^\circ\alpha - \alpha|$ is infinitesimal. □

Theorem 4.4. *Let $^\star\mathfrak{R} = \langle {}^\star\mathbb{R}, \leq, +, \cdot, 0, 1 \rangle$ be a proper ordered field extension of the reals. Then there exists a positive infinitesimal.*
  Proof. Because $^\star\mathfrak{R}$ is a proper extension, let α be an element $^\star\mathbb{R} - \mathbb{R}$. Then $\alpha \neq 0$, and thus $|\alpha| > 0$. There are two possible cases:

(i) $|\alpha| > r$ for each r in $\mathbb{R}^+$, and
(ii) there exist s and t in $\mathbb{R}^+$ such that $s < |\alpha| < t$.

Suppose (i). Then using properties of totally ordered fields, it easily follows that $0 < |\alpha^{-1}| < r^{-1}$ for each r in $\mathbb{R}^+$, that is, $|\alpha^{-1}|$ is a positive infinitesimal. Suppose (ii). Then $|{}^\circ\alpha - \alpha| > 0$, and by Theorem 4.3, $|{}^\circ\alpha - \alpha|$ is infinitesimal. □

Theorem 4.5. *Let $\langle {}^{*}\mathbb{R}, \leq, +, \cdot, 0, 1\rangle$ be an ordered field extension of the reals and α and β be finite elements of $\mathbb{R}$ such that $\alpha < \beta$. Then ${}^{\circ}\alpha \leq {}^{\circ}\beta$.*
    Proof. Because for each r in $\mathbb{R}^{+}$,

$$|{}^{\circ}\alpha - \alpha| < r \ \text{ and } \ |{}^{\circ}\beta - \beta| < r\,,$$

it follows that for each r in $\mathbb{R}^{+}$,

$$ {}^{\circ}\beta - {}^{\circ}\alpha + 2r = ({}^{\circ}\beta + r) + (-{}^{\circ}\alpha + r) > \beta - \alpha > 0\,. $$

That is, for each r in $\mathbb{R}^{+}$,

$$ {}^{\circ}\beta - {}^{\circ}\alpha + 2r > 0\,. $$

Thus ${}^{\circ}\beta - {}^{\circ}\alpha \geq 0$. Therefore ${}^{\circ}\alpha \leq {}^{\circ}\beta$. $\square$

Theorem 4.6. *Let $\langle {}^{*}\mathbb{R}, \leq, +, \cdot, 0, 1\rangle$ be an ordered field extension of the reals and α and β be finite elements of ${}^{*}\mathbb{R}$. Then $\alpha + \beta$ is a finite element of ${}^{*}\mathbb{R}$ and ${}^{\circ}(\alpha + \beta) = {}^{\circ}\alpha + {}^{\circ}\beta$.*
    Proof. It is immediate from Definition 4.3 that $\alpha + \beta$ is finite. By Theorem 4.3, for each r in $\mathbb{R}^{+}$,

$$ |{}^{\circ}\alpha - \alpha| < r\,, \ |{}^{\circ}\beta - \beta| < r\,, \ \text{ and } \ |{}^{\circ}(\alpha + \beta) - (\alpha + \beta)| < r\,. $$

Thus, for each r in $\mathbb{R}^{+}$,

$$ |{}^{\circ}(\alpha + \beta) - ({}^{\circ}\alpha + {}^{\circ}\beta)| < 3r\,. $$

Therefore, ${}^{\circ}(\alpha + \beta) = {}^{\circ}\alpha + {}^{\circ}\beta$. $\square$

Theorem 4.7. *Let $\langle {}^{*}\mathbb{R}, \leq, +, \cdot, 0, 1\rangle$ be a proper ordered field extension of the reals, α and α_1 be infinitesimal elements of ${}^{*}\mathbb{R}$ and β and β_1 be finite elements of ${}^{*}\mathbb{R}$. Then the following five statements are true:*

(1) $\alpha + \alpha_1$ is infinitesimal.
(2) $\alpha \cdot \beta$ is infinitesimal.
(3) If β and β_1 are not infinitesimal, then $\beta \cdot \beta_1$ is finite and not infinitesimal.
(4) $\beta + \beta_1$ is finite.
(5) If β is not infinitesimal, then β^{-1} is finite and not infinitesimal.

    Proof. Left to reader. $\square$

4.1.2   *Extended probability functions*

This book considers several kinds of probability functions. The usual concept of "finitely additive probability function" in the literature will be called a "traditional probability function" to distinguish it from other kinds of probability functions.

Definition 4.5 (traditional probability function). $\mathbb{P}$ is said to be a *traditional probability function* if and only if $\mathbb{P}$ is a function from a boolean algebra of events $\mathfrak{X} = \langle \mathcal{X}, \cup, \cap, -, X, \varnothing \rangle$ into the closed unit interval $[0,1]$ of the reals such that

- $\mathbb{P}(\varnothing) = 0$ and $\mathbb{P}(X) = 1$, and
- *finite additivity:* for all A and B in $\mathcal{X}$, if $A \cap B = \varnothing$, then $\mathbb{P}(A \cup B) = \mathbb{P}(A) + \mathbb{P}(B)$.

If, in addition, $\mathfrak{X}$ is σ-additive and for each pairwise disjoint sequence of disjoint sets in $\mathcal{X}$, $A_1, \ldots, A_i, \ldots$, $\mathbb{P}(\bigcup_{i=1}^{\infty} A_i) = \sum_{i=1}^{\infty} \mathbb{P}(A_i)$, then $\mathbb{P}$ is said to be σ-additive (or *countably additive*).   □

Definition 4.6 (extended probability function). Let $\mathfrak{L} = \langle L, \leq, \sqcup, \sqcap, 1, 0 \rangle$ be a lattice.

Elements a and b of L are said to be *disjoint* if and only if $a \sqcap b = \varnothing$.

$\mathbb{P}$ is said to be an *extended probability function* on $\mathfrak{L}$ if and only if for all a and b in L,

- $\mathbb{P}$ is a function from L into the closed interval of the extended reals $^\star[0,1]$,
- $\mathbb{P}(1) = 1$ and $\mathbb{P}(0) = 0$,
- *finite additivity:* if a and b are disjoint, then $\mathbb{P}(a \sqcup b) = \mathbb{P}(a) + \mathbb{P}(b)$, and
- *monotonicity:* If $a < b$ then $\mathbb{P}(a) < \mathbb{P}(b)$.   □

Note the two important differences between extended probability functions and probability functions:

(1) Extended probability functions may have some non-real values, and
(2) each extended probability function $\mathbb{P}$ satisfies monotonicity, and thus it must satisfy the condition, $\mathbb{P}(a) = 0$ iff $a = 0$.

A traditional probability function $\mathbb{Q}$ can have elements b in its domain such that $b \neq 0$ and $\mathbb{Q}(b) = 0$. As discussed earlier, such situations have as an unfortunate consequence that there may be elements that have a possibility of occurring but have the same probability of occurring as the impossible element 0. This kind of circumstance is eliminated for extended probability functions by monotonicity and allowing for events with infinitesimal probabilities.

Definition 4.7 ($°\mathbb{P}$). Let $\mathfrak{L} = \langle L, \leq, \sqcup, \sqcap, 1, 0 \rangle$ be a lattice and $\mathbb{P}$ an extended probability function on $\mathfrak{L}$. Then, by definition, $°\mathbb{P}$ is the function on L such that for all a in L, $°\mathbb{P}(a)$ is the standard part of $\mathbb{P}(a)$ (Definition 4.4), that is,

$$°\mathbb{P}(a) = °[\mathbb{P}(a)]. \quad \square$$

The following theorem is immediate.

Theorem 4.8. *Suppose $\mathfrak{L} = \langle L, \sqcup, \sqcap, 1, 0 \rangle$ is a lattice and $\mathbb{P}$ is an extended probability function on $\mathfrak{L}$. Then $°\mathbb{P}$ is a traditional probability function.* $\quad \square$

Note that if $\mathbb{P}(a)$ is infinitesimal, then $°\mathbb{P}(a) = 0$.

4.1.3 *L-probabilistic lattices*

For the concept of "extended probability function" to be a productive one, the underlying lattice must be sufficiently rich in disjoint elements in order for finite additivity to have any analytic power. There are many lattices that do not contain such a rich set of disjoint elements, including ones that occur naturally in real world applications. The following definition provides a more general and useful concept of "probability" for lattices that do not depend on having a sufficiently rich set of disjoint elements.

Definition 4.8 (L-probability function, L-probabilistic lattice). Let $\mathfrak{K} = \langle K, \leq, \sqcup, \sqcap, 1, 0 \rangle$ be a lattice. Then $\mathbb{P}$ is said to be a *L-probability function* on $\mathfrak{K}$ if and only if $\mathbb{P}$ is an extended probability function on $\mathfrak{K}$ that satisfies the following condition:
Lattice finite additivity: for all a and b in K,

$$\mathbb{P}(a) + \mathbb{P}(b) = \mathbb{P}(a \sqcup b) + \mathbb{P}(a \sqcap b). \quad \square$$

$\mathfrak{K}$ is said to be *L-probabilistic* if and only if there exists a L-probability function on it. $\quad \square$

Note that lattice finite additivity implies the finite additivity condition for traditional probability functions.

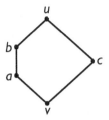

Fig. 4.1 N_5 lattice

The following theorem shows that boolean algebras are examples of L-probabilistic lattices.

Theorem 4.9. *Each extended probability function on a boolean lattice of events is an L-probability function.*

   Proof. Suppose $\mathfrak{X} = \langle \mathcal{X}, \cup, \cap, -, X, \varnothing \rangle$ is a boolean lattice of events, A and B are arbitrary elements of $\mathcal{X}$, and $\mathbb{P}$ is an extended probability function on $\mathfrak{X}$. Because

$$A = (A - B) \cup (A \cap B), \ B = (B - A) \cup (A \cap B),$$

$$\text{and } \ A \cup B = (A - B) \cup (B - A) \cup (A \cap B),$$

it follows from the finite additivity of $\mathbb{P}$ that

$$\mathbb{P}(A) + \mathbb{P}(B) = [\mathbb{P}(A - B) + \mathbb{P}(A \cap B) + \mathbb{P}(B - A)] + \mathbb{P}(A \cap B)$$
$$= \mathbb{P}(A \cup B) + \mathbb{P}(A \cap B). \quad \square$$

Theorem 4.10. *Suppose $\mathfrak{L} = \langle L, \leq, \sqcup, \sqcap, \frown, 1, 0 \rangle$ is a L-probabilistic lattice. Then the following two statements are equivalent:*

(1) $\mathfrak{L}$ is uniquely complemented.
(2) $\mathfrak{L}$ is boolean.

   Proof. Let $\mathbb{P}$ be a L-probability function on $\mathfrak{L}$.
   Suppose Statement 1. Because $\mathfrak{L}$ is complemented, to show that $\mathfrak{L}$ is boolean, it needs to only be shown that $\mathfrak{L}$ is distributive. By Theorems 2.11 and 2.12, to show distributivity it is sufficient that $\mathfrak{L}$ has no N_5 and M_3 sublattices. This is done by contradiction.
   Suppose $\mathfrak{L}$ has a N_5 sublattice, say with distinct elements, a, b, c, u, v with $v < a < b < u$, $u = a \sqcup c = b \sqcup c$, and $v = a \sqcap c = b \sqcap c$ (see Figure 4.1). Then

$$\mathbb{P}(a) + \mathbb{P}(c) = \mathbb{P}(a \sqcup c) + \mathbb{P}(a \sqcap c) = \mathbb{P}(u) + \mathbb{P}(v) = \mathbb{P}(b \sqcup c) + \mathbb{P}(b \sqcap c) = \mathbb{P}(b) + \mathbb{P}(c),$$

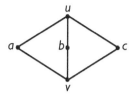

Fig. 4.2 M_3 lattice

and thus $\mathbb{P}(a) = \mathbb{P}(b)$, which is impossible since $a < b$.

Suppose $\mathfrak{L}$ has a M_3 sublattice, say with distinct elements a, b, c, u, and v, where

$$u = a \sqcup b = a \sqcup c = b \sqcup c \quad \text{and} \quad v = a \sqcap b = a \sqcap c = b \sqcap c.$$

(See Figure 4.2.) Because $\mathfrak{L}$ is uniquely complemented, it follows from Lemmas 2.3 and 2.4 that all relative complements of a in $[v, u]$ have the form $v \sqcup (\neg a \sqcap u)$, and thus, because b and c are relative complements of a in $[v, u]$, it must be the case that $b = v \sqcup (\neg a \sqcap u) = c$, which is impossible, because by hypothesis $b \neq c$.

The above shows that Statement 1 implies Statement 2. Statement 2 implies Statement 1 by Theorem 2.18. $\square$

The following theorem shows that for lattices, L-probabilistic implies modular.

Theorem 4.11. *Suppose $\mathfrak{L} = \langle L, \leq, \sqcup, \sqcap, 1, 0 \rangle$ is an L-probabilistic lattice. Then $\mathfrak{L}$ is modular.*

Proof. The proof is essentially a subpart of the proof of Theorem 4.10: By Theorem 2.11, modularity immediately follows if $\mathfrak{L}$ has no N_5 sublattice. Suppose $\mathfrak{L}$ had an N_5 sublattice. A contradiction will be shown. Let a, b, and c be elements of the N_5 sublattice such that

$$a < b, \; b \sqcap c = a \sqcap c < a, \quad \text{and} \quad b < b \sqcup c = a \sqcup c.$$

Then by the definition of L-probability function,

$$\mathbb{P}(a) < \mathbb{P}(b) \tag{4.1}$$

and

$$\mathbb{P}(a) + \mathbb{P}(c) = \mathbb{P}(a \sqcup c) + \mathbb{P}(a \sqcap c) = \mathbb{P}(b \sqcup c) + \mathbb{P}(b \sqcap c) = \mathbb{P}(b) + \mathbb{P}(c),$$

and thus $\mathbb{P}(a) = \mathbb{P}(b)$, contrary to Equation 4.1. $\square$

The following theorem shows how the pseudo complementation of sublattices and together with the existence of an L-probability function yields distributivity.

Theorem 4.12. *Let $\mathcal{L} = \langle L, \sqcup, \sqcap, 1, 0 \rangle$ be a pseudo complemented lattice. Then the following two statements are equivalent.*

(1) $\mathcal{L}$ is distributive.

(2) $\mathcal{L}$ is L-probabilistic and each finite sublattice of $\mathcal{L}$ is pseudo complemented.

Proof. Assume Statement 1. Then, because $\mathcal{L}$ is distributive, it follows from Theorem 4.15 (shown later) that $\mathcal{L}$ is L-probabilistic. Let $\mathcal{L}' = \langle L', \sqcup, \sqcap, u, v \rangle$ be a finite sublattice of $\mathcal{L}$ with unit element u and zero element v, and let a be an arbitrary element of L'. Let

$$C = \{ c \,|\, c \in L' \text{ and } c \sqcap a = v \}.$$

Because L' is a finite set, C is a finite subset of L', say $C = \{c_1, \ldots, c_n\}$. Then, because L' is a lattice, $\bigsqcup_{i=1}^{n} C_i$ is in L'. Because $\mathcal{L}$ is distributive,

$$a \sqcap \bigsqcup_{i=1}^{n} C_i = (a \sqcap c_1) \sqcup \cdots \sqcup (a \sqcap c_i) \sqcup \cdots \sqcup (a \sqcap c_n) = v \sqcup \cdots \sqcup v \sqcup \cdots \sqcup v = v,$$

from which it easily follows that $\bigsqcup_{i=1}^{n} C_i$ is the pseudo complement of a in the lattice $\mathcal{L}'$. This shows Statement 2.

Assume Statement 2. Because $\mathcal{L}$ is L-probabilistic, it follows from Theorems 4.11 and 2.12 that it needs only be shown that $\mathcal{L}$ has no M_3 sublattice. It is immediate that each M_3 lattice is not pseudo complemented, contradicting the hypothesis of Statement 2 that all finite sublattices of $\mathcal{L}$ are pseudo complemented. $\square$

4.2 Dutch Books

Many theorists use The Dutch Book Theorem as a means of justifying traditional probability theory as a rational theory of belief. While there is much merit and usefulness in this, the Dutch Book Theorem—as described below—its use of a pricing mechanism for specifying probabilities and its characterization of the domain of uncertainty as a boolean algebra unnecessarily limits its power and applicability as a rational basis for traditional probability theory. This chapter provides an alternative approach without these limitations, resulting in a stronger, more general, and mathematically and philosophically interesting version of the Theorem. Some of the proofs use methods from linear algebra and the compactness theorem of mathematical logic. These are provided in a separate section at the end of the chapter and in a footnote.

"Degrees of belief" is one way to compare uncertainties inherent in events. Degrees of belief are usually measured using nonnegative real numbers, with a higher number corresponding to a higher degree of belief. The null event, $\varnothing$, has the lowest degree of belief, usually assigned the number 0, and the certain event the highest degree of belief, usually assigned the number 1. The more uncertainty an event has, the smaller its degree of belief and the number assigned to it. De Finetti (1972) presented an argument that is often interpreted by statisticians, economists, and philosophers as demonstrating that traditional probability functions on boolean event space are the only rational approach to the measurement of degrees of belief for uncertain events. His and related arguments are called "Dutch Book Theorems".

Their hypotheses require a participant, let's call him *Bettor,* to price gambles on events from a boolean algebra of events $\mathfrak{X} = \langle \mathcal{X}, \cup, \cap, -, X, \varnothing \rangle$. The gambles have the following form: Pay \$1 if A occurs and \$0 if A does not occur, where A is in $\mathcal{X}$. Such a gamble, which is called an *event gamble,* is denoted by G_A. It is assumed that for each A in $\mathcal{X}$, such a G_A exists and Bettor has to buy or sell each G_A, in any number, at a price, $\mathsf{P}(G_A)$, that is specified by Bettor. For example, if the Bettor buys 3 G_A gambles, then he pays \$3$\mathsf{P}(G_A)$ and receives \$3 if A occurs and \$0 if A does not occur. It is assumed that $\mathsf{P}(G_A) > 0$ for each $A \neq \varnothing$ in $\mathcal{X}$. (Matters can be reformulated so that this assumption is not needed.)

Another participant, let's call her *Arbitrageur,* buys and sells gambles from Bettor trying to make a gain. Arbitrageur is said to *make a Dutch Book on Bettor* if she can buy gambles from Bettor at Bettor's prices and sell gambles to Bettor at Bettor's prices so these transactions produce for Arbitrageur a net monetary gain, no matter the state of the world. In other words, Arbitrageur makes a Dutch Book on Bettor if there is a series of prices for bets that Bettor will accept such that Arbitrageur will receive more from the gambles she purchased than she will pay out from the gambles she sold. Bettor's prices are called *coherent* if and only if no Dutch Book can be made on him.

Define $\mathbb{P}$ on $\mathcal{X}$ as follows: For each A in $\mathcal{X}$,

$$\mathbb{P}(A) = \mathsf{P}(G_A). \tag{4.2}$$

The function $\mathbb{P}$ on $\mathcal{X}$ in Equation 4.2 is called the *price induced probability function on* $\mathcal{X}$. The *Dutch Book Theorem* states that *Bettor's prices are coherent if and only if* $\mathbb{P}$ *is a traditional probability function on* $\mathfrak{X}$. There are various formulations and proofs of the Dutch Book Theorem in the literature, and a proof of the Theorem will not be given in this chapter. Instead, a related, but more general approach, is pursued.

As an example of a situation without coherence, consider the following case: A and B are events in $\mathcal{X}$, where $A \subset B$, Bettor gives G_A and G_B the same price, and Arbitrageur buys G_B from Bettor and sells G_A and G_{B-A} to Bettor. Then the Arbitrageur has a profit of $\mathsf{P}(G_{B-A})$ no matter what state of the world occurs. Note that in this case the Arbitrageur can purchase G_{B-A}, because $\mathfrak{X}$ is a boolean event space, which requires $B - A$ to be an element $\mathcal{X}$.

The above version of the Dutch Book Theorem provides a good argument for rationality of traditional probability. It is applicable for establishing rationality in a wide variety of situations. However, the literature on human decisions under uncertainty provides examples where assumptions about its underlying boolean algebra of events and pricing mechanism are inadequate. For example, bets might be placed on events whose occurrences/non-occurrences cannot be definitely determined and thus the gambles cannot be decided.

The assumptions involving the pricing of gambles restrict in several ways the generality of the Dutch Book Theorem. First, requiring the Bettor to buy or sell any number of gambles at the same price is unrealistic, because the Bettor's utility function u on gambles may not be linear with respect to a number of gambles, that is, it may be the case that $u(n\ G_A) \neq nu(G_A)$ for some positive integer n. Perhaps less important economically, but not less relevant philosophically, is that the combination of defining a Dutch Book in terms of a pricing function on the positive real numbers (e.g., the dollar dimension) and the Arbitrageur making a positive profit may force the Bettor, in cases with an infinite number of events, to assign the same price—and therefore the same probability—to events that he believes have a definite chance of occurring, because the dollar scale does not permit nonzero infinitesimals. For example, consider the following situation: The events are a boolean algebra $\mathcal{B}$ of subevents of the real closed interval $[0,1]$, where $G_{[a,b]}$ for each closed interval $[a,b]$, $0 \leq a < b \leq 1$, is in $\mathcal{B}$ and priced by the Better at $\$(b - a)$, and the gamble $G_{\{c\}}$ for each c in $[0,1]$ is in $\mathcal{B}$ and is priced at $\$0$. Then, for example, the price of $G_{[0,.5] \cup \{.6\}} =$ the price of $G_{[0,.5]}$. Dutch Book arguments usually avoid situations like this, where nonempty events have price \$0, by assuming the underlying event space is finite. However, they are better avoided through a more subtle formulation of "Dutch Book", where the Arbitrageur may not always be guaranteed a monetary profit, but can be in a position of being guaranteed no loss while having a chance (possibly infinitesimal) of winning \$1 or more.

4.3 Preference Books

These issues are removed by reformulating "Dutch Book" in terms of preference instead of price.[2] The idea is that the Bettor has a preference ordering $\precsim$ on gambles. This is more general than prices, because prices can be used to define a preference ordering on gambles by,

$$G_A \precsim G_B \text{ iff } \mathsf{P}(G_A) \leq \mathsf{P}(G_B).$$

In this case, $\precsim$ is *transitive* ($G_A \precsim G_B$ and $G_B \precsim G_C$ implies $G_A \precsim G_C$), *connected* (sometimes called *complete*: either $G_A \precsim G_B$ or $G_B \precsim G_A$ holds), and *reflexive* ($G_A \precsim G_A$). Transitivity and reflexivity are rationality conditions. In the preference version of the Dutch Book Theorem developed below, transitivity is derived instead of being assumed. The rationality of connectedness has been challenged in the literature. In the preference version of the Dutch Book Theorem it is neither assumed nor derived, that is, it is not needed. Reflexivity is a natural rationality condition. It is assumed in the theory developed below.

The key idea involved in the subtle version of the Theorem—where the Arbitrageur is guaranteed no loss but has a chance of winning \$1 or more—is that making a Dutch Book consists of the Arbitrageur buying a set $\mathcal{B}$ of tickets costing \$$b$ and selling a set $\mathcal{S}$ of tickets for \$$s$ so that for each state of the world σ,

$$s - b > 0 \text{ and } \#_\sigma \mathcal{B} \geq \#_\sigma \mathcal{S} \tag{4.3}$$

and for some state of the world τ,

$$\#_\tau \mathcal{B} \geq \#_\tau \mathcal{S},$$

where $\#_\sigma \mathcal{B}$ is the number tickets in $\mathcal{B}$ having σ as a winning state, and similarly $\#_\sigma \mathcal{S}$ is the number tickets in $\mathcal{S}$ having σ as a winning state. The below preference version of the Dutch Book Theorem employs a related idea based on pairwise trades of gambles instead of bulk purchases and sales

---

[2]Pedersen (2014) has a general version of the Dutch Book Argument based on preference over random variables, and he also deals with events of probability 0 in a manner similar to the one presented in this chapter.

of gambles to produce a cleaner theorem with stronger rationality results. Using the terminology inherent Equation 4.3, the preference version changes Equation 4.3 to the following:

> The Arbitrageur cannot lose any money (that is, $s - b \geq 0$) but has an opportunity to make money (that is, $\#_\sigma \mathcal{B} \geq \#_\sigma \mathcal{S}$ for all states of the world σ, and for some state of the world τ, $\#_\tau \mathcal{B} > \#_\tau \mathcal{S}$).

The Dutch Book Theorem, like traditional probability theory, assumes a boolean algebra of events. Nowhere in the literature have I encountered a justification of this assumption as a necessary condition for a rational approach to degrees of belief. Theorems presented in this chapter imply that distributive algebra of events are just as good as boolean algebra of events for rationality considerations along the line of Dutch Book arguments.

Throughout the rest of this chapter, the following definitions and conventions will generally hold, unless stated explicitly otherwise.

- $\precsim$ is a reflexive relation on the set of ticket gambles based on events from the distributive algebra of events $\mathfrak{D} = \langle \mathcal{X}, \cup, \cap, -, X, \varnothing \rangle$. In some situations $\mathfrak{D}$ will be specialized to a boolean algebra of events, and it will be explicitly noted when this is done. (It should also be noted that $\precsim$ induces a reflexive relation on $\mathcal{X}$, which with a minor abuse of notation is also denoted by $\precsim$, where $A \precsim B$ iff $G_A \precsim G_B$. This use of $\precsim$ as a relation on $\mathcal{X}$ is used later in the chapter.)
- $G_A \precsim G_B$ stands for "The Bettor is willing to trade G_A for G_B".
- By definition, $G_A \prec G_B$ iff $[G_A \precsim G_B$ and not: $G_B \precsim G_A]$.
- By definition, $G_A \sim G_B$ iff $[G_A \precsim G_B$ and $G_B \precsim G_A]$.
- A relation of the form $G_A \prec G_B$ is called a $\prec$-*strict inequality*.
- A relation of the form $G_A \sim G_B$ is called an $\sim$-*equivalence*.
- $\mathbb{P}$ is said to be an *extended probability representation* for $\precsim$ if and only if $\mathbb{P}$ is an extended probability function on $\mathfrak{D}$ and for all A and B in $\mathcal{X}$, (*i*) if $G_A \prec G_B$ then $\mathbb{P}(A) < \mathbb{P}(B)$, and (*ii*) if $G_A \sim G_B$ then $\mathbb{P}(A) = \mathbb{P}(B)$.
- $\mathcal{T} = \{T_i \mid i \in I\}$ is said to be an *indexed set of strict inequalities or equivalences* if and only if I is a nonempty finite set such that for each i in I, T_i is either a $\prec$-strict inequality or a $\sim$-equivalence. Repeats of strict inequalities or equivalences are allowed.
- Let $\mathcal{T} = \{T_i \mid i \in I\}$ be an indexed set of strict inequalities or equivalences. By definition, for each x in X,

$$\#_{L,\mathcal{T}}(x) = \text{the number of } i \text{ such that } x \text{ is an element of the left of } T_i$$

and

$\#_{R,\mathcal{T}}(x) =$ the number of i such that x is an element of the right of T_i .

We are now in a position to apply a version of the key idea to obtain a different and more general version of the Dutch Book Theorem. This new version makes an apparently weaker assumption in the sense that there may be fewer kinds of potential Dutch Books to be avoided than before. However, this yields a stronger result concerning the existence of an extended probability representation. Once the existence of an extended probability function $\mathbb{P}$ on $\mathfrak{D}$ has been established, it can be shown that all potential Dutch Books are avoided, that is, the apparently weaker assumption turns out to be not logically weaker than before, because it can be shown that all potential Dutch Books are avoided when the extended probability function $\mathbb{P}$ is used as a price function (instead of a traditional probability function) in the Dutch Book Theorem.

To make terminology less confusing, matters will be formulated in terms of "Preference Books" instead of Dutch Books.

The Bettor is said to be *free from $\precsim$-Preference Books* if and only if for each indexed set of strict inequalities or equivalences $\mathcal{T}$, if for each x in X,

$$\#_{L,\mathcal{T}}(x) = \#_{R,\mathcal{T}}(x)$$

then $\mathcal{T}$ consists only of equivalences. The following theorem follows from Theorem 4.18 of Subsection 4.5.3.

Theorem 4.13. *Let $\precsim$ be a reflexive relation on gambles based on events from a boolean algebra of events. Then the Bettor is free from $\precsim$-Preference Books if and only if there exists an extended probability representation for $\precsim$.*

Theorem 4.13 follows from Narens (1974). It also follows for finite boolean algebras by Scott (1964). Narens extended Scott's results through a straightforward application of an ultraproduct construction. Instead of having $\precsim$ be defined on event gambles, Scott and Narens had it defined directly on events. This practice is followed in the next section, where Theorem 4.13 is generalized to distributive algebras of events.

The condition "being free from $\precsim$-Preference Books" is a version of a different set of conditions due to Scott (1964). Scott called his version "the finite cancellation axioms", and this is the standard terminology for these conditions in the measurement theory literature. Besides having $\precsim$

as a relation on events instead of gambles, Scott had a different way of formulating the finite cancellation axioms than the way presented here.[3]

Scott and Narens did not propose the finite cancellation axioms as a version of the Dutch Book Theorem. Instead they employed them as a qualitative correlate corresponding to the quantitative operation of addition $+$ on a subset of real numbers. They used this result to obtain representation theorems in both probabilistic and non-probabilistic situations where the operation of addition played a central role.

4.4 Finite Cancellation Axioms

Define $\precsim$ on $\mathcal{X}$ as follows: For all A and B in $\mathcal{X}$,

$$A \precsim B \text{ iff } G_A \precsim G_B.$$

Then "$\prec$-strict inequality", "$\sim$-equivalence", "index set of strict inequalities or equivalences," "$\#_{L,T}(x)$", "$\#_{R,T}(x)$", etc., can be defined in terms of this $\precsim$ relation on $\mathcal{X}$, by applying previous definitions to $\precsim$ (considered as a relation on $\mathcal{X}$). When this is done for "the Bettor is free from Preference Books", it is identical to what Narens (1974) calls "$\precsim$ satisfying the finite cancellation axioms" (Definition 4.9 below). In this way, the substance of Theorem 4.13 can be completely reformulated in terms of a distributive algebra of events $\mathfrak{X}$ and the relation $\precsim$ on $\mathcal{X}$ to yield the generalization Theorem 4.14 below of Theorem 4.13.

Definition 4.9. Let $\mathfrak{D} = \langle \mathcal{X}, \cup, \cap, X, \varnothing \rangle$ be a distributive algebra of events, $\precsim$ be a reflexive relation $\mathcal{X}$, and $T = \{T_i \mid i \in I\}$ be a nonempty finite indexed set consisting of $\precsim$-strict inequalities or $\precsim$-equivalences. For each x in X, let

- $\#_{L,\mathcal{T}}(x)$ be the number of i in I such that x is an element of the left side of T_i

and

- $\#_{R,\mathcal{T}}(x)$ be the number of i in I such that x is an element of the right side of T_i.

---

[3]Scott's result is one of the fundamental representation theorems of modern measurement theory. At about the same time his article was submitted for publication, Tversky had independently proved the same result, but Scott's paper reached the *Journal of Mathematical Psychology* before Tversky submitted his to that journal.

Then the *finite cancellation axioms* is said to hold for $\precsim$ if and only if for each x in X and each nonempty indexed set $\mathcal{S}$ consisting of $\precsim$-strict inequalities or $\precsim$-equivalences, if $\#_{L,\mathcal{S}}(x) = \#_{R,\mathcal{S}}(x)$ then $\mathcal{S}$ consists of $\precsim$-equivalences. □

Theorem 4.14. *Let $\mathfrak{D} = \langle \mathcal{X}, \cup, \cap, X, \varnothing \rangle$ be a distributive algebra of events and $\precsim$ be a reflexive relation $\mathcal{X}$. Then the following two statements are equivalent:*

(1) There exists an L-probability function $\mathbb{P}$ on $\mathcal{X}$ such that for all A and B in $\mathcal{X}$,

- *if $A \prec B$ then $\mathbb{P}(A) < \mathbb{P}(B)$,*
- *if $A \sim B$ then $\mathbb{P}(A) = \mathbb{P}(B)$, and*
- *if $A \subset B$ then $\mathbb{P}(A) < \mathbb{P}(B)$.*

(2) $\precsim$ satisfies the finite cancellation axioms.

The proof of Theorem 4.14 employs methods from linear algebra and the compactness theorem of logic. It is given in Subsection 4.5.3 as Theorem 4.18. □

Note that because $\precsim$ is only assumed to be reflexive in Theorem 4.14, there may be no A and B such that $A \prec B$. Thus there may be no interesting relationship between $\prec$ and $\subset$.

Theorem 4.15. *Let $\mathfrak{Y}$ be distributive algebra of events with domain $\mathcal{Y}$. Then the following two statements hold:*

(1) There exists an L-probability function on $\mathfrak{Y}$.
(2) If $\mathfrak{Y}$ is boolean, there exists an extended probability function on $\mathfrak{Y}$.

Proof. Define $\precsim$ on $\mathcal{Y}$ as follows: For all A and B in $\mathcal{Y}$,

$$A \precsim B \text{ iff } A = B.$$

Then, because for this case

$$\precsim = \sim,$$

$\precsim$ satisfies the finite cancellation axioms, because $\precsim = \sim$ and all $\sim$-equivalences are of the form, $A = A$. Thus by Theorem 4.14 there exists an L-probability function $\mathbb{P}$ on $\mathfrak{Y}$. Because each L-probability function on a boolean algebra of events is an extended probability function, it follows that if $\mathfrak{Y}$ is boolean, then $\mathbb{P}$ is an extended probability function. □

Nikodym (1960) showed Statement 2 by a different method. Statement 1 generalizes his result to distributive lattices.

Definition 4.10 (extension of an L-probability function). Let $\mathfrak{D} = \langle \mathcal{X}, \cup, \cap, X, \varnothing \rangle$ be a distributive algebra of events, $\mathbb{P}$ be an L-probability function on $\mathfrak{D}$, and $\mathfrak{B} = \langle \mathcal{B}, \cup, \cap, -, X, \varnothing \rangle$ be a boolean algebra of events such that $\mathcal{X} \subseteq \mathcal{B}$. Then $\mathbb{P}'$ is said to be *an extension of* $\mathbb{P}$ *that is an extended probability function on* $\mathfrak{B}$ if and only if $\mathbb{P} \subseteq \mathbb{P}'$ and $\mathbb{P}'$ is an extended probability function on $\mathfrak{B}$. $\square$

Theorem 4.16. *Let* $\mathfrak{D} = \langle \mathcal{X}, \cup, \cap, X, \varnothing \rangle$ *be a distributive algebra of events and* $\mathbb{P}$ *be an L-probability function on* $\mathfrak{D}$. *Then there exists an extension* $\mathbb{P}'$ *of* $\mathbb{P}$ *that is an extended probability function on* $\langle \wp(X), \cup, \cap, -, X, \varnothing \rangle$.

The proof of Theorem 4.16 employs the compactness theorem of logic. It is given in Subsection 4.5.3 as Theorem 4.19. $\square$

Tarski (1930) showed that each finitely additive probability function on a boolean algebra of events could be extended to a finitely additive probability function on any larger boolean algebra of events with the same domain. He allowed his probability functions to have nonempty sets of probability 0. Theorem 4.16 states a stronger result: (i) The domain of the probability function is a distributive algebra of events instead of a boolean algebra of events; (ii) the probability function has values in a totally ordered field extension of reals; and (iii) the probability function has the empty event as the only event that has probability 0.

From a Dutch Book rationality perspective, Theorem 4.16 justifies expanding rationality to L-probability functions on distributive lattices: By the Dutch Book Theorem, the probability function $\mathbb{P}'$ in Theorem 4.16 is a rational assignment of probabilities, and therefore no Dutch Book can be against the pricing function described by $\mathbb{P}'$. Because in that theorem the domain of $\mathbb{P}'$ is a superset of the domain of the probability function $\mathbb{P}$ defined on a distributive algebra of events, it follows that no Dutch Book can be made against the pricing function described by $\mathbb{P}$.

However, it should be noted that a pricing function $\mathbb{Q}$ on a lattice $\mathfrak{L}$ that gives rise to an L-probability function on $\mathfrak{L}$ is not enough to guarantee its rationality from a Dutch Book perspective: Consider the M_3 lattice of events $\mathfrak{L}$ in Figure 4.3. Let

$$\mathbb{Q}(v) = 0, \quad \mathbb{Q}(a) = \mathbb{Q}(b) = \mathbb{Q}(c) = \frac{1}{2}, \quad \mathbb{Q}(u) = 1.$$

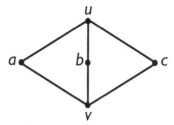

Fig. 4.3 A M_3 lattice of events where $u = \{a, b, c\}$ and $v = \varnothing$.

Then $\mathbb{Q}$ is an L-probability function on $\mathfrak{L}$. A Dutch Book can be made against $\mathbb{Q}$ by selling each of a, b, and c at $\$\frac{1}{2}$ and buying X at $\$1$. Somewhat similarly, having a pricing function $\mathbb{Q}'$ on $\mathfrak{L}$ that is free from Dutch Books does not guarantee that it is an L-probability function: For example, let

$$\mathbb{Q}'(v) = 0, \quad \mathbb{Q}'(a) = \mathbb{Q}'(b) = \mathbb{Q}'(c) = \frac{1}{3}, \quad \mathbb{Q}(u) = 1.$$

The above discussion shows that an L-probability function on a lattice $\mathfrak{L}$ is not a sufficient guarantee that the assignment of probabilities is rational: More has to be assumed about the algebraic structure of the lattice. The discussion also suggests that a reasonable condition for this is that $\mathfrak{L}$ has no M_3 sublattices. Because the existence of an L-probability function on $\mathfrak{L}$ implies $\mathfrak{L}$ is modular (Theorem 4.11), it follows from $\mathfrak{L}$ having no M_3 sublattices that it is distributive (Theorem 2.12). This makes distributive algebras of events natural candidates for rational assignment of probabilities.

This view is also consistent with the few articles in the literature that employ intuitionistic logic instead of classical logic as a foundation for a rational theory of probability. Weatherson (2003), has argued that intuitionistic logic, in the guise of a finite distributive algebra of events with operators of pseudo complementation (for intuitionistic negation) and relative pseudo complementation (for intutionistic implication), provides a better basis and insights for a Dutch Book theorem than classical formulations of the Dutch Book Theorem that employ a boolean algebra of events. His intuitionistic approach allows for the use of Kripke frames to justify additional rationality considerations. Because his algebra is finite and distributive, he could have started with a finite distributive algebra of events as his event space, because such an algebra of events is easily shown to be pseudo complemented and relatively pseudo complemented.

4.5   Additional Theorems and Proofs

This section provides missing proofs of Theorems 4.14 and 4.16. These proofs use the Compactness Theorem of logic.

4.5.1   *Preliminary lemma*

This subsection provides a preliminary lemma that is useful in proofs involving infinite distributive algebras of events.

Lemma 4.1. *Suppose $\mathbb{P}$ is an L-probability function on a finite distributive algebra of events* $\mathfrak{X} = \langle \mathcal{X}, \cup, \cap, X, \varnothing \rangle$ *and* $\mathfrak{B} = \langle \mathcal{B}, \cup, \cap, -, X, \varnothing \rangle$ *is the boolean algebra of events that is generated by* $\mathcal{X} - \varnothing$. *Then* $\mathfrak{B}$ *is finite and there exists an extended probability function* $\mathbb{P}'$ *on* $\mathfrak{B}$ *that extends* $\mathbb{P}$ *from* $\mathcal{X}$ *to* $\mathcal{B}$.

   Proof. Because $\mathcal{X}$ is finite, it follows from Theorem 2.31 that $\mathcal{B}$ is finite. Let $\mathbb{Q}$ be a largest function such that

(1) $\mathcal{D}$ is the domain of $\mathbb{Q}$,

(2) $\mathcal{D} \subseteq \mathcal{B}$,

(3) $\mathfrak{D} = \langle \mathcal{D}, \cup, \cap, X, \varnothing \rangle$ is a distributive lattice of events,

(4) $\mathbb{P} \subseteq \mathbb{Q}$, and

(5) $\mathbb{Q}$ is an L-probability function on $\mathfrak{D}$.

$\mathbb{Q}$ exists, because $\mathbb{P}$ satisfies (1) to (5) with $\mathcal{X} = \mathcal{D}$. The lemma then follows by showing $\mathcal{D} = \mathcal{B}$. This will be done by contradiction.
   Suppose $\mathcal{D} \neq \mathcal{B}$. Because $\mathcal{X}$ is finite,

- let $\mathcal{A}$ be the nonempty set of atoms of $\mathfrak{B}$, and
- let $\mathcal{A}_{\mathcal{D}}$ be the set of atoms of $\mathfrak{B}$ that are in $\mathcal{D}$.

If $\mathcal{A} = \mathcal{A}_{\mathcal{D}}$, then, using Theorem 2.29, $\mathcal{A}_{\mathcal{D}}$ generates $\mathfrak{B}$. Therefore $\mathfrak{X} = \mathfrak{B}$, because each element b of $\mathfrak{B}$ is a finite union of elements of $\mathcal{A}$ and therefore a finite union of elements of $\mathcal{A}_{\mathcal{D}}$. This shows that $\mathcal{B} \subseteq \mathcal{D}$. But $\mathcal{D} \subseteq \mathcal{B}$, because $\mathfrak{B}$ is the boolean algebra of events generated by $\mathfrak{D}$. This contradicts the hypothesis that $\mathcal{D} \neq \mathcal{B}$. Thus $\mathcal{A} - \mathcal{A}_{\mathcal{D}} \neq \varnothing$. Thus let

$$E \in (\mathcal{A} - \mathcal{A}_{\mathcal{D}}) \quad \text{and} \quad \mathcal{E} = \{ D \,|\, D \in \mathcal{D} \quad \text{and} \quad E \subseteq D \}.$$

Because $\mathfrak{B}$ is finite, it follows that $\mathcal{D}$ is finite and therefore $\mathcal{E}$ is finite. Thus let

$$F = \bigcap \mathcal{E}. \tag{4.4}$$

Then $F \in \mathcal{D}$, because $\mathcal{D}$ is closed under $\cap$. Also, because $X \in \mathcal{E}$, it follows that $\mathcal{E} \neq \varnothing$ and thus, by the definition of F,

$$E \subseteq F. \tag{4.5}$$

Because $E \in (\mathcal{A} - \mathcal{A}_\mathcal{D})$ and $F \in \mathcal{D}$, it follows from Theorem 2.29 that $E \neq F$. Thus $E \subset F$, and because E is an atom of $\mathcal{B}$, $\varnothing \subset E \subset F$. Thus,

$$\mathbb{Q}(F) > 0. \tag{4.6}$$

Note that

$$D \cap \{E\} = \varnothing. \tag{4.7}$$

Let $\mathcal{D}'$ be the closure of $\mathcal{D} \cup \{E\}$ under $\cup$, that is, let $\mathcal{D}'$ be the smallest set such that

- $\mathcal{D} \cup \{E\} \subseteq \mathcal{D}'$, and
- for all A and B, if A and B are $\mathcal{D}'$, then $(A \cup B) \in \mathcal{D}'$.

The following shows that $\mathfrak{D}' = \langle \mathcal{D}', \cup, \cap, X, \varnothing \rangle$ is a distributive subalgebra of events of $\mathfrak{B}$:

Because $\mathcal{D}$ and $\{E\}$ are subsets of $\mathcal{B}$, it follows from the definition of $\mathcal{D}'$ and that $\mathfrak{B}$ is a boolean algebra of events that $\mathcal{D}' \subseteq \mathcal{B}$. It follows from the definition of $\mathfrak{D}$ that X and $\varnothing$ are in $\mathcal{D}$, and therefore they are in $\mathcal{D}'$. By the definition of $\mathcal{D}'$, it is closed under $\cup$. The following shows that $\mathcal{D}'$ is closed under $\cap$ and therefore is a distributive subalgebra of events of $\mathfrak{B}$: All events G in $\mathcal{D}'$ has one of the following two forms:

(*i*) $G \in D$ or $G \in \{E\}$.
(*ii*) $G = H \cup K$, where $H \in D$ and $K \in \{E\}$.

Let G_1 and G_2 be arbitrary elements of $\mathcal{D}'$. It then follows from (*i*) and (*ii*), Equation 4.7, and the distributivity of $\mathfrak{B}$ that

$$(G_1 \cap G_2) \in D \quad \text{or} \quad (G_1 \cap G_2) \in \{E\}.$$

Thus $(G_1 \cap G_2) \in \mathcal{D}'$.

The following defines an L-probability function $\mathbb{Q}'$ for $\mathfrak{D}'$ that extends $\mathbb{Q}$, contradicting the choice of $\mathbb{Q}$. Let $\mathbb{Q}'$ on be the binary relation on $\mathcal{D}'$ such that for each G in $\mathcal{D}'$,

(1') $\mathbb{Q}'(G) = \mathbb{Q}(G)$ if $G \in D$,
(2') $\mathbb{Q}'(E) = \frac{1}{2}\mathbb{Q}(F)$, (where F is as defined in Equation 4.4), and
(3') $\mathbb{Q}'(D \cup E) = \mathbb{Q}'(D) + \mathbb{Q}'(E)$.

Then, because $E \notin \mathcal{D}$, $\mathbb{Q}'$ is a function on $\mathcal{D}'$ that extends $\mathbb{Q}$. Because $\mathbb{Q}$ is into $^*[0,1]$, it follows that $\mathbb{Q}'$ is into $^*[01]$. Because $\mathcal{D} \cap \{E\} = \varnothing$ and $\varnothing \in \mathcal{D}$, it follows from $(3')$ that for each G in $\mathcal{D}'$,

$$\begin{aligned}
\mathbb{Q}'(G \cup E) &= \mathbb{Q}'(G) + \mathbb{Q}'(E) \\
&= \mathbb{Q}'(G) + \mathbb{Q}'(E) - \mathbb{Q}'(\varnothing) \\
&= \mathbb{Q}'(G) + \mathbb{Q}'(E) - \mathbb{Q}'(G \cap E),
\end{aligned}$$

showing that $\mathbb{Q}'$ is L-additive on $\mathcal{D}'$. To show $\mathbb{Q}'$ is monotonic, suppose A and B are arbitrary elements of $\mathcal{D}'$ such that $A \subset B$. Let G and H in $\mathcal{D}$ be such that

$$A = G \cup E \quad \text{and} \quad B = H \cup E.$$

Then $G \subset H$. Thus, by $(2')$,

$$\mathbb{Q}'(A) = \mathbb{Q}'(G) + \mathbb{Q}'(E) < \mathbb{Q}'(H) + \mathbb{Q}'(E) = \mathbb{Q}'(B),$$

which contradicts the assumed maximality of $\mathbb{Q}$. $\square$

4.5.2 *Proof of Scott's Theorem*

The proof of Scott's Theorem below employs the following well-known lemma concerning the simultaneous solution to a finite number of strict inequalities and equations in n unknowns.

Lemma 4.2. *The system of the following k strict real inequalities in the unknowns $x_1, \ldots, x_n$,*

$$\begin{aligned}
a_1^1 x_1 + \cdots + a_n^1 x_n &> 0 \\
a_1^2 x_1 + \cdots + a_n^2 x_n &> 0 \\
&\vdots \\
a_1^k x_1 + \cdots + a_n^k x_n &> 0,
\end{aligned}$$

and $m - k$ real equations,

$$\begin{aligned}
a_1^{k+1} x_1 + \cdots + a_n^{k+1} x_n &= 0 \\
a_1^{k+2} x_1 + \cdots + a_n^{k+2} x_n &= 0 \\
&\vdots \\
a_1^m x_1 + \cdots + a_n^m x_n &= 0,
\end{aligned}$$

either has a simultaneous real solution, $p_1 = x_1$, ..., $p_n = x_n$, or there exist non-negative real numbers $r_1, \ldots, r_k$, not all of which are 0, and real numbers $r_{k+1}, \ldots, r_m$ such that for $j = 1, \ldots, n$,

$$\sum_{i=1}^{m} r_i a_j^i = 0.$$

Furthermore, if a_j^i is rational for $j = 1, \ldots, n$ and $i = 1, \ldots, m$, then $r_1, \ldots, r_k$ can be chosen to be rational.

Proof. A nice proof and discussion of this well-known lemma is given in Chapter 2 of Krantz, Luce, Suppes, and Tversky (1971). □

Theorem 4.17 (Scott's Theorem). *Suppose X is finite,*

$$\mathfrak{B} = \langle \mathcal{X}, \cup, \cap, -, X, \varnothing \rangle$$

is a boolean algebra of events, $\precsim$ is a reflexive relation on $\mathcal{X}$ such that $\varnothing \prec X$, and $\precsim$ satisfies the finite cancellation axioms. Then there exists a traditional probability function $\mathbb{P}$ on $\mathcal{X}$ such that for all A and B in $\mathcal{X}$,

- *if $A \prec B$ then $\mathbb{P}(A) < \mathbb{P}(B)$,*
- *if $A \sim B$ then $\mathbb{P}(A) = \mathbb{P}(B)$,*

and

- *if $A \subset B$ then $\mathbb{P}(A) < \mathbb{P}(B)$.*

Proof. Let Γ be the set of $\precsim$-strict inequalities that hold between elements of $\mathcal{X}$ and Σ be the set of $\precsim$-equivalences that hold between elements of $\mathcal{X}$. Because X is finite, Γ and Σ are finite. Let n be the number of elements in X and $X = \{x_1, \ldots, x_n\}$. For each A in $\mathcal{X}$, let $\overline{A}$ be the following vector in $\mathbb{R}^n$:

$$\overline{A} = (a_1, \ldots, a_n),$$

where for $i = 1, \ldots, n$,

$$a_i = \begin{cases} 1 \text{ if } x_i \in A \\ 0 \text{ if } x_i \notin A. \end{cases}$$

For each γ in $\Gamma \cup \Sigma$, if γ is $A \prec B$ or is $A \sim B$, let

$$\overline{\gamma} = \overline{A} - \overline{B}.$$

It will be shown by contradiction that there exists a vector c in $\mathbb{R}^n$ such that for all γ,

$$\text{if } \gamma \in \Gamma \text{ then } c \cdot \overline{\gamma} > 0 \tag{4.8}$$

and

$$\text{if } \gamma \in \Sigma \text{ then } c \cdot \overline{\gamma} = 0. \qquad (4.9)$$

Suppose that there does not exist c in $\mathbb{R}^n$ that satisfy Equations 4.8 and 4.9 for all γ in $\Gamma \cup \Sigma$. Because $\varnothing \prec X$ and $X \sim X$, it follows that $\Gamma \neq \varnothing$ and $\Sigma \neq \varnothing$. Let

$$\Gamma = \{\gamma^1, \dots, \gamma^k\} \text{ and } \Sigma = \{\gamma^{k+1}, \dots, \gamma^p\}$$

be listings of the elements of Γ and Σ. Then by Lemma 4.2, let $r_1, \dots, r_p$ be rational numbers such that

- $r_1, \dots, r_k$ are nonnegative,
- not all of $r_1, \dots, r_k$ are 0,
- and

$$\sum_{i=1}^{p} r_i \overline{\gamma}^i_j = 0 \text{ for } j = 1, \dots, n, \qquad (4.10)$$

where of course $\overline{\gamma}^i_j$ is the jth coordinate of the vector $\overline{\gamma}^i$.

In order to make use of the hypothesis that the finite cancellation axioms hold in this proof, an analog to Equation 4.10 is needed in which r_i are nonnegative integers. To create such an analog, define s_i and λ^i for $i = 1, \dots, p$ as follows:

- If $i = 1, \dots, k$, then $s_i = r_i$ and $\lambda^i = \gamma^i$.
- If $i = k+1, \dots, p$ and r_i is nonnegative, then $s_i = r_i$ and $\lambda^i = \gamma^i$.
- If $i = k+1, \dots, p$ and r_i is negative and A and B are such that $\gamma^i = A \sim B$, then $s_i = -r_i$ and $\lambda^i = B \sim A$.

The result of this is that for $i = 1, \dots, p$, s_i is nonnegative and $s_i \overline{\lambda}^i_j = r_i \overline{\gamma}^i_j$. Thus

$$\sum_{i=1}^{p} s_i \overline{\lambda}^i_j = \sum_{i=1}^{p} r_i \overline{\gamma}^i_j = 0 \quad \text{for } j = 1, \dots, n.$$

Multiplication of both sides of the equation

$$\sum_{i=1}^{p} s_i \overline{\lambda}^i_j = 0$$

by the common denominator w of the positive rational numbers s_i, $i = 1, \dots, p$, yields

$$\sum_{i=1}^{p} t_i \overline{\lambda}^i_j = 0 \quad \text{for } j = 1, \dots, n, \qquad (4.11)$$

where $t_i = w s_i$, $i = 1, \ldots, p$, are nonnegative integers.

Let

$$\Lambda' = \{\lambda^i \mid 1 \le i \le p\}.$$

For each λ in Λ' let, by definition, for nonnegative integers m, $m\{\lambda\}$ be an (indexed) set of strict inequalities or equivalences that contains exactly m copies of λ. Let

$$\Lambda - t_1\{\lambda^1\} \cup \cdots \cup l_p\{\lambda^p\},$$

where $t_1, \ldots, t_p$ are as in Equation 4.11. Let x_j be an arbitrary element of X. Then it follows from Equation 4.11 that the number of times x_j occurs on the left side of an element of Λ of the form λ^i, $\#_{L,\Lambda}(x_j)$ is the same as the number of times it occurs on the right side of some element of Λ, $\#_{R,\Lambda}(x_j)$. Thus for each x in X,

$$\#_{L,\Lambda}(x) = \#_{R,\Lambda}(x).$$

Thus, because Λ is a finite set and the finite cancellation axioms hold, Λ is a set of equivalences. However, because $\varnothing \prec X$ is Γ and therefore is in Λ, Λ contains a strict inequality, which contradicts that Λ is a set equivalences.

The previous argument by contradiction shows that Equations 4.8 and 4.9 hold. It then follows that for each A and B in $\mathcal{G}$,

$$\text{if } A \prec B \text{ then } c \cdot \overline{A} < c \cdot \overline{B} \tag{4.12}$$

and

$$\text{if } A \sim B \text{ then } c \cdot \overline{A} = c \cdot \overline{B}. \tag{4.13}$$

Then, because $\overline{\varnothing} = (0, \ldots, 0)$, it follows that $c \cdot \overline{\varnothing} = 0$. Because $\varnothing \prec X$, it follows from Equation 4.12 that $0 = c \cdot \overline{\varnothing} < c \cdot \overline{X}$.

Define the function $\mathbb{P}$ on $\mathcal{X}$ as follows: For each A in $\mathcal{X}$,

$$\mathbb{P}(A) = \frac{c \cdot \overline{A}}{c \cdot \overline{X}}.$$

Then $\mathbb{P}(X) = 1$, $\mathbb{P}(\varnothing) = 0$, and for all A and B in $\mathcal{X}$,

$$\text{if } A \prec B, \text{ then } c \cdot \overline{A} < c \cdot \overline{B}, \text{ and thus } \mathbb{P}(A) < \mathbb{P}(B),$$

and

$$\text{if } A \sim B, \text{ then } c \cdot \overline{A} = c \cdot \overline{B}, \text{ and thus } \mathbb{P}(A) = \mathbb{P}(B).$$

Suppose A and B are arbitrary elements of $\mathcal{X}$ such that $A \cap B = \varnothing$. Then

$$\overline{A \cup B} = \overline{A} + \overline{B},$$

and thus

$$\mathbb{P}(A \cup B) = \frac{c \cdot \overline{(A \cup B)}}{c \cdot \overline{X}} = \frac{c \cdot (\overline{A} + \overline{B})}{c \cdot \overline{X}} = \frac{c \cdot \overline{A} + c \cdot \overline{B}}{c \cdot \overline{X}} = \mathbb{P}(A) + \mathbb{P}(B).$$

The above shows that $\mathbb{P}$ is a traditional probability function satisfying the statement of the theorem. $\square$

4.5.3   *Proofs using the compactness theorem of logic*

Theorem 4.18. *Let* $\mathfrak{D} = \langle \mathcal{X}, \cup, \cap, -, X, \varnothing \rangle$ *be a distributive algebra of events and* $\precsim$ *be a reflexive relation* $\mathcal{X}$. *Then the following two statements are equivalent:*

(1) There exists an L-probability function $\mathbb{P}$ *on* $\mathcal{X}$ *such that for all A and B in* $\mathcal{X}$,

- *if* $A \prec B$ *then* $\mathbb{P}(A) < \mathbb{P}(B)$, *and*
- *if* $A \sim B$ *then* $\mathbb{P}(A) = \mathbb{P}(B)$.

(2) $\precsim$ *satisfies the finite cancellation axioms.*

Proof. Assume Statement 1. Let k be an arbitrary positive integer, $\mathcal{T} = \{T_1, \ldots, T_k\}$, where for i, $i = 1, \ldots, k$, T_i is either a $\precsim$-equivalence of elements of $\mathcal{X}$ or a $\precsim$-strict inequality of elements of $\mathcal{X}$. Suppose for each x in X, $\#_{L,\mathcal{T}}(x) = \#_{R,\mathcal{T}}(x)$. It needs only to be shown that $\mathcal{T}$ does not contain a $\precsim$-strict inequality. Suppose $\mathcal{T}$ contained a $\precsim$-strict inequality. A contradiction will be shown.

Let $\mathcal{D}$ be the set of events A such that A occurs on the left or right side of an element of $\mathcal{T}$. Because $\mathcal{D}$ is finite and $\mathcal{D} \subseteq \wp(D)$ and $\langle \wp(D), \cup, \cap, -, D, \varnothing \rangle$ is a boolean algebra of events, it follows by Theorem 2.31 that there is a finite boolean algebra of events

$$\mathfrak{E} = \langle \mathcal{E}, \cup, \cap, -, X, \varnothing \rangle$$

such that $\mathcal{D} \subseteq \mathcal{E}$. By Theorems 2.29 and 2.30, the set of atoms $\mathcal{A}$ of $\mathcal{E}$ is finite and

- each element of $\mathcal{A}$ is nonempty,
- for each F in $\mathcal{A}$ and each G in $\mathcal{E}$, if $\varnothing \subset G \subseteq F$, then $F = G$,
- each element of $\mathcal{E}$ different from $\varnothing$ is a finite union of elements of $\mathcal{A}$, and
- the elements of $\mathcal{A}$ are disjoint.

For $i = 1, \ldots, k$, let

$$T_i = A_i \lhd_i B_i, \text{ where } \lhd_i \text{ is either } \sim \text{ or } \prec.$$

Then, by Statement 1, for $i = 1, \ldots, k$,

$$\text{if } \lhd_i \text{ is } \prec, \text{ then } \mathbb{P}(A_i) < \mathbb{P}(B_i),$$

and

$$\text{if } \lhd_i \text{ is } \sim, \text{ then } \mathbb{P}(A_i) = \mathbb{P}(B_i).$$

Therefore, because by hypothesis $\mathcal{T}$ contains a $\precsim$-strict inequality,

$$\sum_{i=1}^{k} \mathbb{P}(A_i) < \sum_{i=1}^{k} \mathbb{P}(B_i).$$

However, because by hypothesis $\mathcal{E}$ is finite, it follows that for each nonempty C in $\mathcal{E}$,

$$\mathbb{P}(C) = \sum_{F \in \mathcal{A} \text{ and } F \subseteq C} \mathbb{P}(F),$$

and thus,

$$\sum_{i=1}^{k} \sum_{F \in \mathcal{A} \text{ and } F \subseteq A_i} \mathbb{P}(F) = \sum_{i=1}^{k} \mathbb{P}(A_i) < \sum_{i=1}^{k} \mathbb{P}(B_i) \qquad (4.14)$$

$$= \sum_{i=1}^{k} \sum_{F \in \mathcal{A} \text{ and } F \subseteq B_i} \mathbb{P}(F).$$

Because the elements of $\mathcal{A}$ are disjoint, the only way for

$$\sum_{j=1}^{k} \mathbb{P}(A_j) < \sum_{i=1}^{k} \mathbb{P}(B_i) \qquad (4.15)$$

is for there to be some element H in $\mathcal{A}$ such that the set

$$\{i \mid 1 \leq i \leq k \ \& \ H \subseteq B_i\}$$

is larger than the set $\{j \mid 1 \leq j \leq k \ \& \ H \subseteq A_j\}$. However, because such a H is nonempty (a property of being in $\mathcal{A}$), it follows that there exist x in H and i, $1 \leq i \leq k$ such that x is an element of the left side of an element of $\mathcal{T}$ fewer times than it an element of the right side of an element of $\mathcal{T}$—that is, $x \in A_i$ for fewer i, $i = 1, \cdots, k$, than $x \in B_i$. Because $H \in \mathcal{A}$ and $\mathcal{A} \subseteq \mathcal{E} \subseteq \mathcal{X}$, this contradicts the assumption

$$\#_{L,\mathcal{T}}(x) = \#_{R,\mathcal{T}}(x).$$

Statement 2 implying Statement 1 will be shown by the Compactness Theorem of logic. Assume Statement 2.

Let L be the first-order language that has the following predicate, operation, and individual constant symbols:

- For describing an ordered field extension of the reals:

  $\mathbf{R}(x)$, $x \leq y$, $x + y = z$, $x \cdot y = z$, and $\boldsymbol{r}$ for each real number r.

- For describing the distributive lattice $\mathfrak{D}$ with domain $\mathcal{X}$ and sure event X:

$$\mathcal{X}, \cup, \cap, -, X, \varnothing, \mathbf{A} \text{ for each } A \text{ in } \mathcal{X}.$$

- For describing the boolean algebra of events $\mathfrak{B} = \langle \mathcal{B}, \cup, \cap, -, X, \varnothing \rangle$ containing $\mathfrak{D}$ as a subalgebra:

$$\mathcal{B}, \cup', \cap', -', \varnothing, \mathbf{B} \text{ for each } B \text{ in } \mathcal{B}.$$

- For describing an extended probability function on $\mathfrak{B}$ that appropriately relates a description of $\precsim$ to the description of $\leq$:

$$x \precsim y \text{ and } P(x) = y.$$

Let:

- Γ be the set of first-order sentences of L formulated in terms of $\mathbf{R}$, $\leq$, $+$, $=$, saying $\mathbf{R}$ is the domain of a totally ordered field and for each r, s, and t in $\mathbb{R}$,

$$\mathbf{R}(\boldsymbol{r}), \quad r \leq s \text{ iff } \boldsymbol{r} \leq \boldsymbol{s},$$
$$r + s = t \text{ iff } \boldsymbol{r} + \boldsymbol{s} = \boldsymbol{t},$$
$$r \cdot s = t \text{ iff } \boldsymbol{r} \cdot \boldsymbol{s} = \boldsymbol{t}.$$

Thus Γ says that $\mathbf{R}$ is the domain of a totally ordered field extension of the reals.

- Δ be the set of first-order sentences of L formulated in terms of $\mathcal{X}$, $\cup$, $\cap$, $-$, X, $\varnothing$, and $\mathbf{A}$ for each A in $\mathcal{X}$ saying $\mathcal{X}$ is the domain of a distributive algebra of events (formulated in the obvious manner in L with sure event denoted by X) and $\mathcal{X}(\mathbf{A})$ for each A in $\mathcal{X}$.

- Σ be the set of first-order sentences of L formulated in terms of $\mathcal{B}$, $\cup'$, $\cap'$, $-'$, X, and $\varnothing$ saying $\mathcal{B}$ is the domain of a boolean algebra of events with sure event X and empty event $\varnothing$ (formulated in the obvious manner in L).

- Π be the set of first-order sentences of L that says

  (i) $\mathcal{X}$ is a subset of $\mathcal{B}$.

  (ii) The distributive algebra of events described in Δ is a subalgebra of the boolean algebra of events described in Σ.

  (iii) $\mathbf{P}$ is an L-probability function of the distributive algebra of events with domain described by $\mathcal{X}$ that is described in Δ.

  (iv) Assume the proper definitions have been given in L for $\prec$, $\sim$, and $<$, and that $\rightarrow$ is the implication symbol of L. Then for all A and B in $\mathcal{X}$,

$$A \prec B \ \rightarrow \ P(A) < P(B) \,,$$

and

$$A \sim B \ \rightarrow \ P(A) = P(B),$$

- Ω be the set of atomic sentences of L that are true about $\mathfrak{D}$, for example, if r, s, and t are in $\mathbb{R}$, then

$$\boldsymbol{r+s=t} \text{ is in } \Omega \text{ iff } r+s=t; \ \boldsymbol{r \leq s} \text{ is in } \Omega \text{ iff } r \leq s; \text{ etc.,}$$

if A, B, and C are in $\mathcal{D}$, then

$$\boldsymbol{A \cup B = C} \text{ is in } \Omega \text{ iff } A \cup B = C; \text{ etc.;}$$

if A and B are in $\mathcal{D}$ and r is in $\mathbb{R}$, then

$$\boldsymbol{A \precsim B} \text{ is in } \Omega \text{ iff } A \precsim B; \text{ and } \boldsymbol{P(A) = r} \text{ iff } \mathbb{P}(A) = r; \text{ etc.;}$$

etc.
- $\Lambda = \Gamma \cup \Delta \cup \Sigma \cup \Pi \cup \Omega \,.$

Let Φ be a finite subset of Λ. Let α be the set of constant symbols occurring in Φ of the form $\boldsymbol{D}$ for D in $\mathcal{X}$. Let $\gamma = \alpha \cup \{\boldsymbol{X, \varnothing}\}$. Let $\mathcal{F} = \{D \,|\, \boldsymbol{D} \in \gamma\}$. Then $\mathcal{F}$ is a finite subset of the domain $\mathcal{X}$ of $\mathfrak{D}$ and thus $\mathcal{F}$ is a finite subset of the boolean algebra of events $\langle \wp(X), \cup, \cap, -, X, \varnothing \rangle$. Therefore, by Theorem 2.31, $\mathcal{B}$ can be chosen so that

$$\mathfrak{B} = \langle \mathcal{B}, \cup, \cap, -, X, \varnothing \rangle$$

is the finite boolean algebra of events generated by $\mathcal{F}$. Define the reflexive relation $\precsim'$ on $\mathcal{B}$ as follows: For all A and B in $\mathcal{B}$,

- if A and B are in $\mathcal{F}$ and $A \precsim B$ then $A \precsim' B$, and
- if A is in $\mathcal{B} - \mathcal{F}$ then $A \sim' A$.

Because, by hypothesis, $\precsim$ satisfies the finite cancellation, it is easy to verify that $\precsim'$ satisfies the finite cancellation axioms. Thus, by Theorem 4.17 (Scott's Theorem), let $\mathbb{P}'$ be a traditional probability function on $\mathfrak{B}$ such that

- if $A \prec B$ then $\mathbb{P}'(A) < \mathbb{P}'(B)$;
- if $A \sim B$ then $\mathbb{P}'(A) = \mathbb{P}'(B)$;

and

- if $A \subset B$ then $\mathbb{P}'(A) < \mathbb{P}'(B)$.

Then all sentences in Φ are true about $\mathfrak{B}$ (interpreted as a distributive algebra of events) under the obvious interpretations of the symbols of L, i.e., with $\mathbf{R}$ interpreted as the set of real numbers, etc., $\boldsymbol{\mathcal{X}}$ and interpreted as $\mathcal{B}$, $\boldsymbol{X}$ and interpreted as X, etc., $\boldsymbol{P}$ interpreted $\mathbb{P}'$, and $\precsim$ interpreted as $\precsim'$.

Therefore, by the Compactness Theorem of logic, let $\mathfrak{M}$ be a model of Λ. Because the sentences in Ω are true about $\mathfrak{D}$, it follows that $\mathfrak{D}$ is isomorphically embeddable in $\mathfrak{M}$. *Therefore, without loss of generality, it is assumed that $\mathfrak{D}$ is a substructure of $\mathfrak{M}$.* For $\mathfrak{M}$, let

- ${}^{\star}\mathfrak{R} = \langle {}^{\star}R, \leq, +, \cdot, 0, 1 \rangle$ be the order extension of the reals that is a model of Γ with the interpretations ${}^{\star}R$ for $\mathbf{R}$, r for $\boldsymbol{r}$ (r in $\mathbb{R}$), etc.;
- ${}^{\star}\mathfrak{D} = \langle {}^{\star}\mathcal{X}, \cup, \cap, -, X, \varnothing \rangle$ be the distributive algebra of events that is a model of Δ with interpretations ${}^{\star}\mathcal{X}$ for $\boldsymbol{\mathcal{X}}$, $\cup$ for $\cup$, $\cap$ for $\cap$, $-$ for $-$, X for $\boldsymbol{X}$, $\varnothing$ for $\varnothing$, and A for $\boldsymbol{A}$ for each A in $\mathcal{X}$;
- ${}^{\star}\precsim$ be the interpretation $\precsim$ and ${}^{\star}\mathbb{P}$ be the interpretation of $\boldsymbol{P}$.

Note that it follows from part (iii) of Π that ${}^{\star}\mathbb{P}$ is an L-probability function from ${}^{\star}\mathcal{X}$ into ${}^{\star}\mathbb{R}$ and that all A and B in ${}^{\star}\mathcal{X}$,

- if $A \; {}^{\star}\!\prec B$ then $\mathbb{P}(A) \; {}^{\star}\!< \mathbb{P}(B)$,

and

- if $A \sim B$ then ${}^{\star}\mathbb{P}(A) = {}^{\star}\mathbb{P}(B)$.

Let $\mathbb{P}$ be the restriction of ${}^{\star}\mathbb{P}$ to the domain $\mathcal{X}$ of $\mathfrak{D}$. Then $\mathbb{P}$ is an L-probability function, and Statement 1 follows. $\quad\square$

Theorem 4.19. *Let*

- $\mathfrak{D} = \langle \mathcal{X}, \cup, \cap, X, \varnothing \rangle$ *be a distributive algebra of events,*
- $\mathfrak{B} = \langle \mathcal{Y}, \cup, \cap, Y, \varnothing \rangle$ *be a boolean algebra of events,*
- $\mathcal{X} \subseteq \mathcal{Y}$, *and*
- *and $\mathbb{P}$ be an L-probability function from $\mathfrak{D}$ into the extended real number system*

$$\mathfrak{R}' = \langle \mathbb{R}', \leq, +, \cdot, 1, 0 \rangle.$$

Then there exists an extension $\mathbb{P}'$ of $\mathbb{P}$ that is an extended probability function on $\mathfrak{B}$.

Proof. The proof of Theorem 4.19 is similar to Theorem 4.18, except $\mathfrak{B}$ is described differently and it extends the language L so that $\mathbb{P}'$ can also be described, and it uses a related but different argument to show the existence of $\mathbb{P}'$.

Let L' be the first-order language that has predicate and operations symbols and individual constant symbols for describing an ordered field extension of the reals,

$$\mathbf{R}(x),\ x \leq y,\ x + y = z,\ x \cdot y = z,\ \boldsymbol{r} \text{ for each } r \text{ in } \mathbb{R}',$$

predicate and operations symbols and individual constant symbols for describing the distributive lattice $\mathfrak{D}$,

$$\mathcal{X},\ \cup,\ \cap,\ -,\ \boldsymbol{X},\ \varnothing,\ \mathbf{A} \text{ for each } A \text{ in } \mathcal{X},$$

predicate and operations symbols and individual constant symbols for describing a boolean algebra of events $\mathfrak{B} = \langle \mathcal{B}, \cup, \cap, -, Y, \varnothing \rangle$ containing $\mathfrak{D}$ as a sublattice,

$$\boldsymbol{\mathcal{B}},\ \cup',\ \cap',\ -',\ \boldsymbol{Y},\ \varnothing,\ \mathbf{B} \text{ for each } B \text{ in } \mathcal{B},$$

a predicate symbol

$$\precsim$$

for describing $\precsim$, and function symbols

$$\boldsymbol{P}(x) = y \text{ and } \boldsymbol{P}'(x) = y$$

for describing, respectively, the L-probability function $\mathbb{P}$ on $\mathfrak{D}$ and the extended probability function $\mathbb{P}'$ on $\mathfrak{B}$.

Let:

- Γ' be the set of first-order sentences of L formulated in terms of $\mathbf{R}$, $\leq$, $+$, $=$, saying $\mathbf{R}$ is the domain of a totally ordered field, and for each r, s, and t in $\mathbb{R}'$,

$$\mathbf{R}(\boldsymbol{r}),\ r \leq s \text{ iff } \boldsymbol{r} \leq \boldsymbol{s}$$
$$r + s = t \text{ iff } \boldsymbol{r} + \boldsymbol{s} = \boldsymbol{t}$$
$$r \cdot s = t \text{ iff } \boldsymbol{r} \cdot \boldsymbol{s} = \boldsymbol{t}.$$

Thus Γ' says that $\mathbf{R}$ is the domain of a totally ordered field extension of $\mathfrak{R}'$, which itself is a totally ordered field extension of the reals. Note that Γ in the proof of Theorem 4.18 only assumes $\mathbf{R}$ is the domain of an ordered field extension of the reals, that is, only assumes for each real number r, $\mathbf{R}(\boldsymbol{r})$.

- Δ be the set of first-order sentences of L formulated in terms of $\mathcal{X}$, $\cup$, $\cap$, $-$, $\boldsymbol{X}$, $\varnothing$, and $\boldsymbol{A}$ for each A in $\mathcal{X}$ saying $\mathcal{X}$ is the domain of a distributive algebra of events (formulated in the obvious manner in L with sure event denoted by $\boldsymbol{X}$) and $\mathcal{X}(\boldsymbol{A})$ for each A in $\mathcal{X}$.
- Σ be the set of first-order sentences of L formulated in terms of $\mathcal{B}$, $\cup'$, $\cap'$, $-'$, $\boldsymbol{Y}$, $\varnothing$, and $\boldsymbol{B}$ for each B in $\mathcal{Y}$ saying $\mathcal{B}$ is the domain of a boolean algebra of events (formulated in the obvious manner in L with sure event denoted by $\boldsymbol{Y}$), and $\mathcal{B}(\boldsymbol{B})$ for each B in $\mathcal{B}$.
- Π' be the set of first-order sentences of L that says

(*i*) $\mathcal{B}(\boldsymbol{C})$ for each C in $\mathcal{X} \cup \mathcal{B}$;

(*ii*) the distributive algebra of events described in Δ is a subalgebra of the boolean algebra of events described in Σ;

(*iii*) $\boldsymbol{P}$ is an L-probability function of the distributive algebra of events with domain described by $\mathcal{X}$ that is described in Δ, and is such that for all A in $\mathcal{X}$ there exists r in $\mathbb{R}'$ such that

$$P(\boldsymbol{A}) = \boldsymbol{r}\,; \tag{4.16}$$

and

(*iv*) $\boldsymbol{P}'$ is an extended probability function of the boolean algebra of events with domain described by $\mathcal{B}$ that is described in Σ and for all A in $\mathcal{X}$,

$$\boldsymbol{P}'(A) = \boldsymbol{P}(A)\,.$$

- Ω' be the set of atomic sentences of L. For example, if r, s, and t are in $\mathbb{R}'$, then

$$\boldsymbol{r} + \boldsymbol{s} = \boldsymbol{t} \text{ is in } \Omega \text{ iff } r + s = t;\ \boldsymbol{r} \le \boldsymbol{s} \text{ is in } \Omega \text{ iff } r \le s;\ \text{etc.},$$

if A, B, and C are in $\mathcal{D}$, then

$$\boldsymbol{A} \cup \boldsymbol{B} = \boldsymbol{C} \text{ is in } \Omega \text{ iff } A \cup B = C;\ \text{etc.};$$

and if E is in $\mathcal{D}$ and r is in $\mathbb{R}$, then

$$\boldsymbol{P}(\boldsymbol{E}) = \boldsymbol{r} \text{ is in } \Omega \text{ iff } \mathbb{P}(E) = r\,;$$

and if F is in $\mathcal{B}$ and r is in $\mathbb{R}'$, then

$$\mathbb{P}'(\boldsymbol{F}) = \boldsymbol{r} \text{ is in } \Omega \text{ iff } \mathbb{P}'(F) = r \text{ etc.};$$

etc.

- $\Lambda' = \Gamma' \cup \Delta \cup \Sigma \cup \Pi'$.

Let Φ be a finite subset of Λ'. Let α be the set of constant symbols occurring in Φ of the form $\boldsymbol{B}$ for B in $\mathcal{B}$. (Note by (i) of Π' that the constant symbols of the form $\boldsymbol{A}$ for A in $\mathcal{X}$ are in α.) Let $\gamma = \alpha \cup \{\boldsymbol{Y}, \varnothing\}$. Let $\mathcal{F} = \{B \mid \boldsymbol{B} \in \gamma\}$ and $\mathcal{G} = \{A \mid \boldsymbol{A} \in \gamma$ and $A \in \mathcal{X}\}$. Then $\mathcal{F}$ is a finite subset of the domain $\mathcal{B}$ of the boolean algebra of events $\mathfrak{B}$, and $\mathcal{G}$ is a finite subset of the domain $\mathcal{X}$ of the algebra of events $\mathfrak{D}$. By Theorem 2.31, let

$$\mathfrak{B}' = \langle \mathcal{B}_1, \cup, \cap, -, X, \varnothing \rangle$$

be the finite boolean algebra of events generated by $\mathcal{F}$, and by Theorem 2.32 let

$$\mathfrak{D}' = \langle \mathcal{D}_1, \cup, \cap, -, X, \varnothing \rangle$$

be the finite distributive algebra of events generated by $\mathcal{G}$. Then $\mathcal{D}_1 \subseteq \mathcal{B}_1$. By (iii) of Π', define the function $\mathbb{Q}$ as follows:

$$\mathbb{Q}(A) = r \quad \text{iff} \quad \text{the sentence } \boldsymbol{P(A)} = \boldsymbol{r} \text{ is in } \Pi'.$$

Then it follows from (i) of Π' that the domain of $\mathbb{Q}$ is $\mathcal{D}_1$ and $\mathbb{Q}$ is an L-probability function on $\mathfrak{D}'$. Thus, by Lemma 4.1, let $\mathbb{Q}'$ be an extended probability function on $\mathfrak{B}'$ that extends $\mathbb{Q}$.

Interpret each symbol $\mathbf{S}$ occurring in a sentence of Φ as follows: If $\mathbf{S}$ occurs in a sentence of Γ', then give $\mathbf{S}$ its obvious interpretation in $\mathfrak{R}'$, for example, if $\mathbf{S} = \mathbf{R}$, then $\mathbf{S}$ is interpreted as $\mathbb{R}'$. Similarly, if $\mathbf{S}$ occurs in a sentence of Δ, then give $\mathbf{S}$ its obvious interpretation in $\mathfrak{D}$, and if $\mathbf{S}$ occurs in a sentence of Σ, then give $\mathbf{S}$ its obvious interpretation in $\mathfrak{B}$. Then it easily follows that these obvious interpretations can be used to form a model in which each sentence of Φ is true. Therefore, by the Compactness Theorem of logic, let $\mathfrak{E}'$ be a model of Λ'. For $\mathfrak{E}'$, let:

- $^{\star}\mathfrak{R} = \langle {}^{\star}R, \leq, +, \cdot, 0, 1 \rangle$ be the totally ordered field that is a model of Γ with the interpretations $^{\star}R$ for $\mathbf{R}$, r for $\boldsymbol{r}$ (r in $\mathbb{R}'$), etc. Note that because $\mathbb{R} \subseteq \mathbb{R}'$ and L has a symbol $\boldsymbol{r}$ for each r in $\mathbb{R}'$, it follows that $\mathbb{R} \subseteq \mathbb{R}' \subseteq {}^{\star}\mathbb{R}$, and thus $^{\star}\mathfrak{R}$ is a totally ordered field extension of the reals.
- $^{\star}\mathfrak{D} = \langle {}^{\star}\mathcal{X}, \cup, \cap, -, X, \varnothing \rangle$ be the distributive algebra of events that is a model of Σ with interpretations $^{\star}\mathcal{X}$ for $\mathcal{X}$, $\cup$ for U, $\cap$ for $\mathsf{\cap}$, $-$ for $-$, X for $\boldsymbol{X}$, $\varnothing$ for $\varnothing$, and A for $\boldsymbol{A}$ for each A in $\mathcal{A}$.
- $^{\star}\mathfrak{B} = \langle {}^{\star}\mathcal{B}, \cup, \cap, -, Y, \varnothing \rangle$ be the distributive algebra of events that is a model of Δ with interpretations $^{\star}\mathcal{B}$ for $\mathcal{B}$, $\cup$ for U', $\cap$ for $\mathsf{\cap}'$, $-$ for $-'$, Y for $\boldsymbol{Y}$, $\varnothing$ for $\varnothing$, and B for $\boldsymbol{B}$ for each B in $\mathcal{B}$.
- $^{\star}P$ be the interpretation of $\boldsymbol{P}$ and $^{\star}P'$ be the interpretation of $\boldsymbol{P}'$.

Then it follows from Π' that $^\star P$ is an L-probability function from $^\star \mathfrak{D}$ into $^\star \mathfrak{R}$, and $^\star P'$ is an extended probability function from $^\star \mathfrak{B}$ into $^\star \mathfrak{R}$ that is an extension of $^\star Q$. It also follows from Π' that

$$\mathbb{P} = \text{ the restriction of } ^\star P' \text{ to } \mathcal{D}\,.$$

Let

$$\mathbb{P}' = \text{ the restriction of } ^\star P' \text{ to } \mathcal{B}\,.$$

Then $\mathbb{P}'$ is an extend probability function on $\mathfrak{B}$ that is an extension of $\mathbb{P}$. $\square$

Chapter 5

Rationality, Heuristics, and Human Judgments of Probability

5.1 Introduction

Science and philosophy consider traditional probability theory to be *the* rational approach for measuring degrees of belief involving uncertainty. The literature contains two main lines of argument for this: the Dutch Book Argument and the subjective expected utility (SEU) model. Both employ the concept of "value" or "utility" to justify the additivity of probabilities over a disjoint union of events. Chapter 4 showed that the Argument generalizes to situations involving extended probability functions as well as to qualitative situations based on preference. Chapter 4 also showed that the mathematical driving force behind the Dutch Book Argument could be abstracted to a set of qualitative preference axioms—the finite cancellation axioms—that provided necessary and sufficient conditions for the existence of an extended probability function that is consistent with the preference ordering.

The Dutch Book Argument assumes an underlying boolean algebra of events. Chapter 4 generalizes the Argument so that it applies to a distributive algebra of events, $\mathfrak{D} = \langle \mathcal{D}, \cup, \cap, D, \varnothing \rangle$. This generalization shows that a boolean algebra of events is not necessary for the existence of a useful and robust probability function. Theorem 4.16 of Chapter 4 shows that each extended probability function on $\mathfrak{D}$ extends to an extended probability function on $\langle \wp(D), \cup, \cap, -, D, \varnothing \rangle$. But generally power-sets are too large for probabilistic, empirical investigation. For this and other reasons, the empirical use of power-set algebras have limited empirical use in science. Instead boolean subalgebras of the power-set algebras are employed.

There is not just one kind of uncertainty, but many. These may differ in the selection of the kind of subalgebra of events that is relevant for the assignment of probabilities. For example, one might be in a situation where instances of an event A are clear and easily observable but instances of its complement are unobservable or vague. In such a situation the kind of uncertainty encountered is qualitatively different than when each instance of A or its complement is clearly observed as an instance of A or clearly observed as an instance of the complement of A. Narens (2005) provides examples of this, where empirical events are closed algebraically under set-theoretic unions and intersections but not under set-theoretic complementation. In such examples, a non-complemented distributive algebra of events is more appropriate for probability assignments than a boolean algebra of events. Chapter 6 argues that there are also situations where orthomodular algebras of events are more appropriate than distributive algebras of events for modeling probabilistic concepts.

A *lottery* has the form,

$$L = (a_1, A_1; \ldots; a_i, A_i; \ldots; a_n, A_n),$$

where a_i is a pure outcome, A_i is an event, and "a_i, A_i" stands for receiving a_i if A_i occurs, and $A_i \cap A_j = \varnothing$ for $i \neq j$. The *Subjective Expected Utility Model*—or *SEU* for short—assumes that the decision maker has a utility function u over outcomes and lotteries and a traditional probability function P over events such that for all lotteries $L = (a_1, A_1; \ldots; a_i, A_i; \ldots; a_n, A_n)$,

$$u(L) = \sum_{i=1}^{n} \frac{\mathsf{P}(A_i)}{\mathsf{P}(A_1) + \cdots + \mathsf{P}(A_n)} \cdot u(a_i). \tag{5.1}$$

Equation 5.1 is called the *SEU Model* for L.

Note that in Equation 5.1,

$$\frac{\mathsf{P}(A_i)}{\mathsf{P}(A_1) + \cdots + \mathsf{P}(A_n)}$$

is the subjective conditional probability of A_i occurring given that $\bigcup_{i=1}^{n} A_i$ has occurred.

The claims for rationality of SEU rest on axiomatizations in which the individual axioms are argued to be rational (e.g., the famous axiomatization of Savage, 1954, or the axiomatization of a conditional form of SEU in Chapter 8 of Krantz, Luce, Suppes, and Tversky, 1971).

Some have proposed that rationality should be evaluated in terms of various constraints the decision maker encounters while making decisions. The literature calls this form of "rationality" *bounded rationality* (Simon, 1957), and its constraints may include cognitive ones, like limitations of memory or the ability to make complicated mathematical calculations, or biological ones, such as the effects of emotion generated by the decision situation. One idea pursued in this chapter is that behaviors that are irrational by the Dutch Book or SEU standards may appear more "rational" when the decision process is modeled in a bounded rational manner using a non-boolean algebra of events.

5.2   Cognitive Heuristics

In making decisions, people use various simplifying assumptions called *heuristics*. This chapter focuses on subjective judgments of probability, particularly in situations where people base their judgments on event instances that come to mind. Such judgments are subject to context and heuristics. Various judgmental heuristics have been studied by the psychologists Kahneman and Tversky. For the purposes of this chapter, the most important heuristic is *availability:*

> There are situations in which people assess the frequency of a class or the probability of an event by the ease with which instances or occurrences can be brought to mind. For example, one may assess the risk of heart attack among middle-aged people by recalling occurrences among one's acquaintances. Similarly, one may evaluate the probability that a given business venture will fail by imagining various difficulties it could encounter. This judgmental heuristic is called availability. Availability is a useful clue for assessing frequency or probability, because instances of large classes are usually recalled better and faster than instances of less frequent classes. However, availability is affected by factors other than frequency and probability. *(Tversky & Kahneman, 1974, p. 1127)*

Another important heuristic in decision theory is *representativeness:*

> For an illustration of judgment by representativeness, consider an individual who has been described by a former neighbor as follows: "Steve is very shy and withdrawn, invariably helpful, but with little interest in people, or in the world of reality. A meek and tidy soul, he has a need for order and structure, and a passion for detail." How do people assess the probability that Steve is engaged in a particular occupation from a list of possibilities (for example, farmer, salesman, airline pilot, librarian, or physician)? How do people order these occupations from most to least likely? In the representativeness heuristic, the probability that Steve is a librarian, for example, is assessed by the degree to which he is representative of, or similar to, the stereotype of a librarian. Indeed, research with problems of this type has shown that people order the occupations in exactly the same way (Tversky and Kahneman, 1974). This approach to judgment of probability leads to serious errors, because similarity, or

representativeness, is not influenced by several factors that should affect judgments of probability. *(Tversky & Kahneman, 1974, p. 1124)*

Narens (2007) sees judgments of frequency and representativeness as arising from different kinds of availability. In his modeling, probability judgments involving frequency are based on the exemplars that come to mind (the *availability of exemplars*), whereas probability judgments involving similarity are based on the exemplifying properties that come to mind (the *availability of exemplifying properties*). This difference is illustrated in the following famous example.

Example Kahneman and Tversky (1982) gave participants the following description. β:

> β: Linda is 31 years old, single outspoken and very bright. She majored in philosophy. As a student she was deeply concerned with the issues of discrimination and social justice, and also participated in anti-nuclear demonstrations.

Participants were asked to rank order the following statements by their probability, using 1 for the most probable and 8 for the least probable. The descriptions denoted by γ and α below are the ones that play the important roles in the discussion presented here.

> Linda is a teacher in elementary school
> Linda works in a bookstore and takes Yoga classes
> Linda is active in the feminist movement
> Linda is a psychiatric social worker
> Linda is a member of the League of Women voters
> γ: Linda is a bank teller
> Linda is an insurance salesperson
> α: Linda is a bank teller and is active in the feminist movement

Over 85% of participants ranked as more probable that Linda was both a bank teller and a feminist (α) than just a bank teller (γ). This is an example of what in the literature is called the *conjunction fallacy*. According to Kahneman and Tversky, it is due to representativeness: "bank teller and is active in the feminist movement" is more a "representative" description of Linda than just "bank teller."

Narens (2007) writes the following about the Linda example:

> Cognitive representations take many forms. They all have in common that they are sets, but the sets can have different kinds

of elements. For representativeness, the elements of the sets are taken to be exemplifying properties. This choice allows for a better modeling of the similarity concept.

...the description of Linda, β, makes available to the participant a set of properties, L, that exemplifies people fulfilling that description. Similarly, the predicate "is a bank teller" makes available a set of properties, T, exemplifying bank tellers, and the predicate "is a bank teller and is active in the feminist movement" makes available a set of properties, TF, exemplifying people who are bank tellers and are active in the feminist movement. β, α, and γ are assumed to have the following cognitive representations [where "$\mathbf{CR}(\beta)$" stands for "the cognitive representation of β"]:

- $\mathbf{CR}(\beta) = L$.
- $\mathbf{CR}(\alpha) = L \cap TF$.
- $\mathbf{CR}(\gamma) = L \cap T$.

The "conjunction fallacy" arises because participants give greater support to $L \cap TF$ than $L \cap T$. Theoretically, this occurs because the properties in $L \cap TF$ are more available to the participant than those in $L \cap T$. Also, because T and TF depend on the availability, it is theoretically most likely that for most participants, $TF - T \neq \varnothing$ and $T - TF \neq \varnothing$.

β does not completely characterize a person (real or fictitious); it only gives some characteristics that a person may have. For modeling frequency judgments, the cognitive representation of β should be interpreted as the set D of exemplars d that (cognitively) satisfies β when d is appropriately substituted for "Linda". For modeling similarity judgments, the cognitive representation of β should be interpreted as the set of properties, L. L may be generated in different ways, for example, as the set of properties common to the elements of D, or as the set of properties cognitively derivable from the description β.

The similarity interpretation of proper noun "Linda" is viewed here as the set properties L. The similarity interpretation of the noun phrase "bank teller" also has a set of properties, T, as its cognitive representation. The similarity interpretation of "Linda is a bank teller," γ, is then $L \cap T$. Note how this differs from the [natural English language] semantical representation of γ: In the [natural English language] semantic representation, (*i*) "Linda" is interpreted as an individual, l, not as a set; (ii) the predicate, "is a bank teller," is interpreted as a set t (i.e., the set of bank tellers); and (iii) the statement, "Linda is a bank teller," is interpreted as the statement $l \in t$. In summary, for similarity judgments involving a propositional description, the cognitive representation relates subject and predicate through set theoretic intersection, while the semantic representation relates them through set theoretic membership.

5.3 Support Theory

Support theory is an approach to understanding and modeling human judgments of probability that was initiated in Tversky and Koehler (1994) and extended by a number of investigators (e.g., Rottenstreich and Tversky, 1997; Idson, Krantz, Osherson, and Bonini, 2001). Tversky and Koehler designed it to explain the often seen puzzling empirical result: (1) *binary complementarity:* For a binary partition of an event, the sum of the judged probabilities of elements of the partition is 1, whereas (2) *subadditivity:* for partitions consisting of three or more elements, the sum of the judged probabilities of the elements is ≥ 1, with > 1 often being observed. This chapter provides a novel foundation for support theory due to Narens (2007) that is based on pseudo complemented distributive algebras of events.

The following example of Redelmeier, Koehler, Liberman, and Tversky (1995) demonstrates subadditivity in a between-subjects design. (Other studies show similar effects for within-subject designs.) They presented the following scenario to a group of 52 expert physicians a Tel Aviv University.

> B. G. is a 67-year-old retired farmer who presents to the emergency department with chest pain of four hours' duration. The diagnosis is acute myocardial infarction. Physical examination shows no evidence of pulmonar edema, hypothension, or mental status changes. His EKG shows ST-ssegment elevation in the anterior leads, but no dysrythmia or heart block. His past medical history is unremarkable. He is admitted to the hospital and treated in the usual manner. Consider the possible outcomes.

Each physician was randomly assigned to evaluate on the following four prognoses for this patient:

- dying during this admission
- surviving this admission but dying within one year

- living for more than one year but less than ten years
- surviving for more than ten years.

Redelmeier, et al. write,

> The average probabilities assigned to these prognoses were 14%, 26%, 55%, and 69%, repsectively. According to standard theory, the probabilities assigned to these outcomes should sum to 100%. In contrast, the average judgments added to 164% (95% confidence interval: 134% to 194%).

What is striking is that the subjects in this example are all expert and well experienced, and they are judging a situation that are like ones they encounter routinely and about which they have to make weighty decisions.

Similar results hold for other expert populations. For example, in an experiment of Fox, Rogers, and Tversky (1996), professional option traders were asked to judge the probability that the closing price of Microsoft stock would fall within a particular interval on a specific future date. The experimenters found subadditivity: E.g., when four disjoint intervals that spanned the set of possible prices were presented for evaluation, the sums of the assigned probabilities were typically about 1.50. They also found binary complementarity: When binary partitions were presented the sums of the assigned probabilities were very close to 1, e.g., .98. Fox and Birke (2002) conducted the following experiment involving possible verdicts in the Jones versus Clinton case.

Example: *Jones Versus Clinton.* 200 practicing attorneys were recruited (median reported experience: 17 years) at a national meeting of the American Bar Association (in November 1997). Of this group, 98% reported that they knew at least "a little" about the sexual harassment allegation made by Paula Jones against President Clinton. At the time of that the survey, the case could have been disposed of by either A, which was an outcome other than a judicial verdict, or B, which was a judicial verdict. Furthermore, outcomes other than a judicial verdict can partition A into

(A1) settlement;
(A2) dismissal as a result of judicial action;
(A3) legislative grant of immunity to Clinton; and
(A4) withdrawal of the claim by Jones.

Each attorney was randomly assigned to judge the probability of one of these six events. The results are as follows:

Median Judged Probabilities for All Events in Study

(A) other than a Judicial verdict	.75
(B) judicial verdict	.20
Binary partition total	**.95**
(B) judicial verdict	.20
(A1) settlement	.85
(A2) dismissal	.25
(A3) immunity	.0
(A4) withdrawal	.19
Five fold partition total	**1.49**

Names for an object are usually an arranged collection of symbols. Logic distinguishes an object from its name. For example, the name "Socrates" is distinguished from the ancient Greek philosopher Socrates. Here this distinction is accomplished through the use of quotation marks. In support theory experiments, participants are presented descriptions of an event in a natural language, for example, English, and are asked to provide a probability judgment for the described event. In support theory experiments descriptions of the same event are often manipulated in manners so that people assign different probabilities to the different descriptions of the event. This gives rise to two different kinds of interpretations of descriptions used by participants: A *linguistic interpretation*, denoted by LI, which interprets the description of the event as the event, and a *cognitive interpretation*, denoted by CI, which interprets the description of the event as the mental representation the participant is using *while making her probabilistic judgment*. In this way, different descriptions of an event have the same linguistic interpretation but can have different cognitive interpretations.

Because of this, support theory views subjects as assigning probabilities to descriptions of events instead of to events. An important type of description used in support theory experiments is an explicit disjunction.

Definition 5.1 (explicit disjunction). A description of the form $(\alpha$ or $\beta)$ where α and β are descriptions that describe nonempty events such that the description $(\alpha$ and $\beta)$ describe the empty event is called an *explicit disjunction*. In the obvious manner, the above concept of "explicit disjunction" extends to three or more events. Thus the phrase "explicit disjunction" is sometimes used to described three or more events, for example, "the explicit disjunction (δ or γ or β)" or "the explicit disjunction [(δ or γ) or β]", etc. □

The *data* for most support theory studies consist of the judged (conditional) probabilities of descriptions of the form: "α occurring rather than β occurring", where (α or β) is an explicit disjunction. "α occurring rather than β occurring" can be viewed as describing the event described by α conditioned on the event described by "α or β", in notation, ($\alpha \mid \alpha$ or β). The linguistic and cognitive interpretations are assumed to be related as follows: For all descriptions α and β of events,

if $\mathsf{LI}(\alpha$ and $\beta)$ describes the empty set, then $\mathsf{CI}(\alpha) \cap \mathsf{CI}(\beta) = \varnothing$.

The basic underlying assumption in this chapter's presentation of support theory is that when making a probability judgment about an explicit disjunction α or β, the participant creates a cognitive representation in which she compares evidence for (the event described by) α occurring with evidence against (the event described by) α occurring. In doing this, she uses in the evaluation the cognitive interpretation $\mathsf{CI}(\alpha)$ for evaluating the evidence for α, and the cognitive interpretation $\mathsf{CC}(\alpha)$ for evaluating the evidence against α. The important consideration in this modeling is that is not necessary that $\mathsf{CC}(\alpha) = \mathsf{CI}(\beta)$, and in fact in the interesting cases,

$$\mathsf{CC}(\alpha) \neq \mathsf{CI}(\beta) \,.$$

This chapter's presentation of support theory consists of four parts.

Part 1. A description of an event α has two kinds of interpretations: a *linguistic interpretation*, $\mathsf{LI}(\alpha)$, and a *cognitive interpretation*, $\mathsf{CI}(\alpha)$. The participant uses her linguistic interpretation to decide whether or not two descriptions are logically equivalent. The theory assumes that the descriptions are chosen so that the participant will be correct in this decision, that is, produce the same result as one would obtain by classical logic applied to the ordinary language interpretation of the description. The cognitive interpretation is used by the participant in her calculation of probabilities of descriptions. These calculations involve judgmental heuristics acting on the cognitive interpretation. The linguistic interpretation is largely based on the semantic meanings of the words used in the description. The cognitive interpretation is based on features and biases that arise when the participant makes a probability judgment, for example, features of instances of a remembered category. Linguistic and cognitive interpretations are very different kinds of interpretations and should be separated by any theory involving probability judgments. Support theory does this. Unfortunately, many studies in the literature fail to make this distinction.

Part 2. The descriptions of interest are descriptions of events. The focus of the theory is on cognitive interpretations. Only the following four minimal assumptions are needed for linguistic interpretations: For all descriptions α and β,

- the linguistic interpretation of α has the logical structure of an event, that is, the linguistic interpretation of α is a set;
- if the participant understands α and β are logically equivalent, then $\mathsf{LI}(\alpha) = \mathsf{LI}(\beta)$;
- if the participant understands that the description (α and β) describes $\varnothing$, then $\mathsf{LI}(\alpha) \cap \mathsf{LI}(\beta) = \varnothing$;
- if (α or β) is an explicit disjunction, then $\mathsf{LI}(\alpha \text{ or } \beta) = \mathsf{LI}(\alpha) \cup \mathsf{LI}(\beta)$.

The following is assumed about the cognitive interpretations: There exists a topological algebra of events $\mathcal{T}$ such that for each description α, $\mathsf{CI}(\alpha) \in \mathcal{T}$.

Part 3. The notation $\mathsf{CI}_\Omega(\delta)$ stands for the statement that δ and Ω are descriptions and that the participant understands that the cognitive interpretation of δ is a subset of her cognitive interpretation of Ω. In this case, δ is called the *target description* of the interpretation, and Ω is called the *conditioning description* of the interpretation. For the case of "α occurring rather than β occurring", the target description is α and the conditioning description is (α or β). Usually, the conditioning description will be understood by context, and thus

$$\mathsf{CI}_\Omega(\delta) \text{ is usually written as } \mathsf{CI}(\delta) \text{ and } \mathsf{CI}_{(\alpha \text{ or } \beta)}(\alpha) \text{ as } \mathsf{CI}(\alpha).$$

Support theory assumes that for each description of a nonempty event δ (conditioned on the description Ω), the participant assigns a nonnegative number, $s[\mathsf{CI}(\delta)]$, called the *support of δ (relative to Ω)*. $s[\mathsf{CI}(\delta)]$ is interpreted as a measure of the evidence in favor of the occurrence of the event described by δ. Support theory assumes the participant judges "α occurring given that Ω occurs", in symbols ($\alpha \mid \Omega$), by the formula,

$$\mathbb{P}(\alpha \mid \Omega) = \frac{s[\mathsf{CI}(\alpha)]}{s[\mathsf{CI}(\alpha)] + s[\mathsf{CC}(\alpha)]},$$

where $\mathsf{CC}(\alpha)$ is a cognitive construction the participant uses during the probability judgment to evaluate the size of items that are not in $\mathsf{Cl}(\alpha)$ but are in $\mathsf{Cl}(\Omega)$. $\mathsf{CC}(\alpha)$ is called the *cognitive complement of α (relative to Ω)*, and it is assume that $\mathsf{CC}(\alpha)$ is an element of the topology $\mathcal{T}$. It will often be the case that $\mathsf{Cl}(\alpha) \cup \mathsf{CC}(\alpha) \neq \Omega$. By assumption, $s(\varnothing) = 0$, and cases where $\mathsf{Cl}(\alpha) = \mathsf{CC}(\alpha) = \varnothing$ are not considered or assumed not to occur in the relevant experiments. In terms of this formulation, Tversky and Koehler (1994) can be viewed as assuming for the case of judging "α occurring rather than β occurring",

$$\mathsf{CC}(\alpha) = \mathsf{Cl}(\beta) \text{ and thus } s[\mathsf{CC}(\alpha)] = s[\mathsf{Cl}(\beta)].$$

For the judging of "α occurring rather than β occurring" this chapter only assumes,

$$s[\mathsf{CC}(\alpha)] \leq s[\mathsf{Cl}(\beta)]. \tag{5.2}$$

This is consistent with ideas presented in Rottenstreich & Tversky (1997) which generalizes Tversky & Koehler (1994) to

$$\mathsf{CC}(\alpha) \subseteq \mathsf{Cl}(\beta) \text{ and thus } s[\mathsf{CC}(\alpha)] \leq s[\mathsf{Cl}(\beta)]. \tag{5.3}$$

Note that by Equation 5.2,

$$\begin{aligned}
&\mathbb{P}(\alpha \mid \alpha \text{ or } \beta) + \mathbb{P}(\beta \mid \beta \text{ or } \alpha) \\
&= \frac{s[\mathsf{Cl}(\alpha)]}{s[\mathsf{Cl}(\alpha)] + s[\mathsf{CC}(\alpha)]} + \frac{s[\mathsf{Cl}(\beta)]}{s[\mathsf{Cl}(\beta)] + s[\mathsf{CC}(\beta)]} \\
&\geq \frac{s[\mathsf{Cl}(\alpha)]}{s[\mathsf{Cl}(\alpha)] + s[\mathsf{Cl}(\beta)]} + \frac{s[\mathsf{Cl}(\beta)]}{s[\mathsf{Cl}(\beta)] + s[\mathsf{Cl}(\alpha)]} = 1,
\end{aligned}$$

and thus binary complementarity is violated if $s[\mathsf{CC}(\alpha)] < s[\mathsf{Cl}(\beta)]$ or if $s[\mathsf{CC}(\beta)] < s[\mathsf{Cl}(\alpha)]$. Rottenstreich & Tversky (1997) provides empirical examples of such violations.

Part 4. (δ or γ) is said to be an *unpacking* of the description α if and only if δ or γ is an explicit disjunction of α in the sense that δ or γ is an explicit disjunction and $\mathsf{Ll}(\delta \text{ or } \gamma) = \mathsf{Ll}(\alpha)$. This chapter's theory assumes the following: Suppose δ or γ is an unpacking of α. Then the following principle holds.

Unpacking Principle: $s[\mathsf{Cl}_\Omega(\alpha)] \leq s[\mathsf{Cl}_\Omega(\delta \text{ or } \gamma)] \leq s[\mathsf{Cl}_\Omega(\delta)] + s[\mathsf{Cl}_\Omega(\gamma)].$

Many experimental paradigms in support theory are designed so that participants produce data of the form,

$$\mathbb{P}(\alpha \mid \alpha \text{ or } \beta),$$

where (α or β) is an explicit disjunction, and also produce data of the forms,

$$\mathbb{P}[\delta \,|\, \delta \text{ or } (\gamma \text{ or } \beta)] \text{ and } \mathbb{P}[\gamma \,|\, \gamma \text{ or } (\delta \text{ or } \beta)] ,$$

where (δ or γ) is an explicit disjunction of α, and (γ or β) and (δ or β) are explicit disjunctions. Support theory, as presented here, assumes the following empirically observed relationship:

$$\mathbb{P}(\alpha \,|\, \alpha \text{ or } \beta) \leq \mathbb{P}[\delta \,|\, \delta \text{ or } (\gamma \text{ or } \beta)] + \mathbb{P}[\gamma \,|\, \gamma \text{ or } (\delta \text{ or } \beta)] . \qquad (5.4)$$

Support theory employs various kinds of cognitive processes for determining the support of a description:

> The support associated with a given [description] is interpreted as a measure of the strength of evidence in favor of this [description] to the judge. The support may be based on objective data (e.g., frequency of homicide in the relevant data) or on a subjective impression mediated by judgmental heuristics, such as representativeness, availability, or anchoring and adjustment (Kahneman, Solvic, and Tversky, 1982). For example, the hypothesis "Bill is an accountant" may be evaluated by the degree to which Bill's personality matches the stereotype of an accountant, and the prediction "An oil spill along the eastern coast before the end of next year" to be assessed by the ease with which similar accidents come to mind. Support may also reflect reasons or arguments recruited by the judge in favor of the hypothesis in question (e.g., if the defendant were guilty, he would not have reported the crime). *(Tversky and Koehler, 1994, p. 549)*

This chapter employs a topology of open sets to model the structural relationships of various "subjective impressions mediated by" the judgmental heuristic of availability described in the above quotation and their influence on probability judgments. The basic idea is that descriptions of events with the same linguistic interpretation produce cognitive events having different levels of availability. For example, the unpacking of α into γ or δ is likely to cause the sum of the number recalled instances of γ and the number recalled instances of δ to be larger than the number of recalled instances of α. $\mathsf{Cl}(\alpha)$, for a description α, is interpreted as the set of *clear instances* of α, that is, as the set of instances i that come to mind that it is clear to the participant, when she is judging the probability of α, that i is an instance of α. An instance that comes to mind that is not clearly an instance of α is called a *poor instance* of α. *Vague instances*—instances that come to mind that are sort of instances of α—and *confusing instances*—instances that appear to the participant and belonging to α and then again not to α—are examples of poor instances. In finding the support of α for a probability judgment, the participant is assumed to ignore poor instances. This is why the support for α is written as $s[\text{Cl}(\alpha)]$. Similarly, $\text{CC}(\alpha)$ consists of clear instances of Ω that come to mind in the judging of the probability of $(\alpha \mid \Omega)$ that are clearly not instances of α.

This chapter interprets $\text{Cl}(\alpha)$ as an open set in a topology, $\mathcal{T}$. $\mathcal{T}$ is defined as follows, where $\mathcal{D}$ is the nonempty set of descriptions under consideration:

$$\mathcal{T} = \text{the topology generated by } \{\text{Cl}(\alpha) \mid \alpha \in \mathcal{D}\} \cup \{\text{CC}(\alpha) \mid \alpha \in \mathcal{D}\}.$$

The universal set of $\mathcal{T}$ is called the *set of clear instances (of $\mathcal{D}$)*, and elements of this universal set are called *clear instances (of $\mathcal{D}$)*. $\mathcal{T}$ is used for characterizing various linkages between interpretations of descriptions. In particular, for the cognitive interpretations of α and its unpacking (δ or γ), it follows that

$$\dot{-}\dot{-}\text{Cl}(\alpha) = \dot{-}\dot{-}\text{Cl}(\delta \text{ or } \gamma) \text{ and } \text{CC}(a) \subseteq \dot{-}\text{Cl}(\alpha), \tag{5.5}$$

where $\dot{-}$ is the operation of pseudo complementation of the underlying topology. *It should be emphasized that it is not assumed that each element of $\mathcal{T}$ is an interpretation of a description in $\mathcal{D}$.* In particular, it is not assumed that $\dot{-}\dot{-}\text{Cl}(\alpha)$ is necessarily the cognitive interpretation of some description in $\mathcal{D}$, and similarly for $\text{Cl}(\alpha) \cap \text{Cl}(\beta)$. (Note that even if ($\alpha$ and β) is a description in $\mathcal{D}$, it does not necessarily follow that $\text{Cl}(\alpha \text{ and } \beta) = \text{Cl}(\alpha) \cap \text{Cl}(\beta)$.)

Availability is related to the ease and number of instances come to mind. When availability is used for evaluating support of a description, generally a recalled set of items will have greater support than its proper subsets. For the Unpacking Principle, this suggests that more clear instances are available through the separate evaluations involving $\text{Cl}(\delta)$ and $\text{Cl}(\gamma)$ of the unpacking of (δ or γ) of α than are available in evaluating α. A stronger version of this is that

$$\text{Cl}(\alpha) \subseteq \text{Cl}(\delta \text{ or } \gamma). \tag{5.6}$$

Note that $\mathbb{P}(\alpha \mid \Omega) \leq \mathbb{P}(\delta \text{ or } \gamma \mid \Omega)$ and Equation 5.5 does not imply Equation 5.6.

Various kinds of instances that are associated with α when the participant is judging $(\alpha \mid \Omega)$ are modeled as boundary elements of $\text{Cl}(\alpha)$. For example, consider an instance i in $\text{Cl}(\Omega)$ that is not a clear instance in the judging of the probability of α but is a clear instance of some unpacking (δ or γ) of α. Because of Equation 5.5, such an instance is in $\dot{-}\dot{-}\alpha$. It is called a *potential clear instance of* α. Thus $[\dot{-}\dot{-}\,\mathsf{Cl}(\alpha)] - \mathsf{Cl}(\alpha)]$ is the set of potential clear instances of α. (Narens, 2007, gives additional topological interpretation to this set by modeling it as a subset of the boundary of $\mathsf{Cl}(\alpha)$.) Theoretically, potential clear instances are responsible for empirical observations of the Unpacking Principle when the availability heuristic plays a primary role in probability judgments. One possible reason for this is that the more specific descriptions of δ and γ of an unpacking of α are likely to bring to mind more instances in $\dot{-}\dot{-}\alpha$ than α.

Some clear instances i of α are *behaviorally ambiguous* in the sense that i is a clear instance of α when, for example, the participant is judging $\mathbb{P}(\alpha\,|\,\alpha$ or $\beta)$ and is clear instance in the later judging of $\mathbb{P}(\beta\,|\,\alpha$ or $\beta)$. An example of this is a participant judging the probability that a randomly selected gem from an exhibition she has just been at is green rather than blue. Because she might be concentrating more on "green" than "blue" in making this judgment, she might exhibit a Whorfian context effect and classify a blue-green gem as a clear instance of "green". Later when judging the probability of a blue rather than a green gem, she might again exhibit a Whorfian context effect and classify the same gem as a clear instance of "blue". Behavioral ambiguity should be contrasted with the clear instance i of $\mathsf{Cl}(\alpha)$ being *cognitively ambiguous* in the sense that it is in $\mathsf{Cl}(\alpha) \cap \mathsf{CC}(\alpha)$. Still other clear instances i of α are *vague* in the sense that there is only weak evidence that i is in $\mathsf{Cl}(\alpha)$ and no evidence or almost no evidence that it is in $\mathsf{CC}(\alpha)$. (Narens, 2007, models behaviorally ambiguous, cognitively ambiguous, and vague instances of α as qualitatively different kinds of topological boundary points of $\mathsf{Cl}(\alpha)$.) Behaviorally ambiguous instances contribute in the evaluations of support, whereas cognitive ambiguous instances and vague instances are poor instances and are ignored in the evaluations of support.

In modeling probability judgments and decision making, the vast majority of researchers employ a boolean algebra of events. This is a special form of a topological event model—one for which the topology is such that each open set is also a closed set. A justification for this choice of event space is almost never provided for these kinds of behavioral applications. The topological modeling of events in this section allows for various features cognitive processing presumably involved in subjective evaluations of probabilities to be part of the structure of the event space. These then get reflected in features of probability assignments. This should be contrasted with approaches that assume a boolean algebra of events and alter the algebraic form of the behavioral probability function to account for various features of the cognitive processing in making probability judgments. The topological event space approach also allows for a richer language and theory for describing the processing. For support theory, this section's topological modeling appears to me to provide a more conceptually rich and rigorous foundation than the approaches of Tversky & Koehler and Rottenstreich & Tversky.

5.4   Rationality

The conditional form of the SEU (subjective expected utility) model for decision making assumes that the decision maker has a preference ordering $\precsim$ on lotteries, a utility function u on pure outcomes, and a traditional probability function P over events such that for all lotteries $L = (a_1, A_1; \ldots; a_i, A_i; \ldots; a_n, A_n)$ and $L' = (a'_1, A'_1; \ldots; a'_j, A'_j; \ldots; a'_m, A'_m)$,

$$L \precsim L' \text{ iff} \tag{5.7}$$

$$\sum_{i=1}^{n} \frac{\mathsf{P}(A_i)}{\mathsf{P}(A_1) + \cdots + \mathsf{P}(A_n)} \cdot u(a_i) \leq \sum_{i=1}^{m} \frac{\mathsf{P}(A'_j)}{\mathsf{P}(A'_1) + \cdots + \mathsf{P}(A'_m)} \cdot u(a_j).$$

(Note that in Equation 5.7,

$$\frac{\mathsf{P}(A_i)}{\mathsf{P}(A_1) + \cdots + \mathsf{P}(A_n)}$$

is the subjective conditional probability of A_i occurring given that $\bigcup_{i=1}^{n} A_i$ has occurred.)

Much experimental research has shown that humans tend to violate SEU in systematic ways. Some in the literature have argued that SEU is a misguided standard for rationality for human decision making. For example, Gilboa, Postlewaite, and Schmeidler (2009) write,

> We reject the view that rationality is a clear-cut, binary notion that can be defined by a simple set of rules or axioms. There are various ingredients to rational choice. Some are of internal coherence, as captured by Savage's axioms. Others have to do with external coherence with data and scientific reasoning. The question we should ask is not whether a particular decision is rational or not, but rather, whether a particular decision is more rational than another. And we should be prepared to have conflicts between the different demands of rationality. When such conflicts arise, compromises are called for. Sometimes we may relax our demands of internal consistency; at other times we may lower our standards of justifications for choices. But the quest for a single set of rules that will universally define the rational choice is misguided.

This section presents a version of bounded rationality due to Narens (2014). It develops a notion of "rationality" for evaluating lotteries in which decision maker changes states while making decisions. This type rationality is called *cognitive rationality*, and the type of rationality inherent in SEU, is called *objective rationality*. They are very similar. They both use the SEU formula,

$$\sum_{i=1}^{n} \frac{\mathsf{P}(A_i)}{\mathsf{P}(A_1) + \cdots + \mathsf{P}(A_n)} \cdot u(a_i), \qquad (5.8)$$

for computing utility of a given lottery. They differ in the computation of utilities across lotteries: Objective rationality requires that across lotteries the same event must be given the same probability; subjective rationality, on the other hand, allows for the possibility that the same event occurring in different lotteries can have different probabilities.

The generalization of SEU presented here is called *DSEU* or *descriptive SEU*. It is based on the following simple framework for the cognitive processing of lotteries: Lotteries are construed to be things in the every day world. These are called *objective lotteries*. Accurate descriptions of them are presented to the decision maker for preference evaluation. The decision maker forms *cognitive representations* of them producing a corresponding set of *subjective lotteries*. These are assumed to retain relevant formal properties of lotteries that he then employs in a *decision process*. For a pair of lotteries, the decision process yields an *intended action* of decision maker to make a preferred lottery choice. This intended action yields an observable preference action in the world on objective lotteries. (See Figure 5.1.) This way viewing the decision process leads naturally to the concepts of *subjective* and *objective rationality*. In both of these concepts, there is a rational consistency between the cognitive representation and preference in the intended action in that they are connected by SEU. This rational consistency is called *subjective coherence*. Objective rationality has a different consistency that relates objective preference actions to objective lotteries through SEU. This consistency is called *objective coherence*. The principle difference between subjective and objective rationality occurs in the cognitive representation. In objective rationality, an event A occurring in objective lotteries is given the same cognitive interpretation across objective lotteries; in subjective rationality, an event A occurring in different lotteries may have different cognitive interpretations in different lotteries. For DSEU, an outcome appearing in different lotteries is always given the same interpretation for both objective and subjective rational-

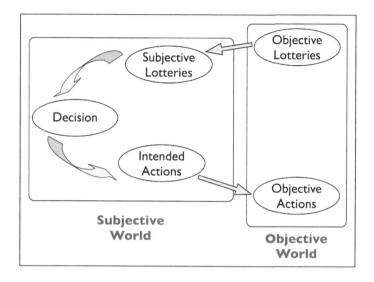

Fig. 5.1 Processing Framework

ity. The possible differences in interpretations in subjective and objective rationality leads to subjective rationality not necessarily having objective coherence. These concepts are illustrated in Figures 5.2 and 5.3.[1]

5.4.1 *Modeling DSEU*

In DSEU, the decision maker is modeled as being in various states. The states of interest are those that influence his cognitive representations of lotteries. These can have a complicated structure. For example, individual outcomes or events in a lottery may determine the state of the decision maker. For example, an outcome that is catastrophic may produce fear in the decision maker, causing him to evaluate the event associated with it in a different manner than he would if that event was associated with, say, the outcome of receiving $10. For the purposes of this section, such within lot-

---

[1]The DSEU framework can be generalized so that subjective rationality can also include situations where the cognitive representation of a lottery outcome can depend on more than just the outcome itself, allowing it to have multiple interpretations in the cognitive representation phase of the cognitive processing and multiple utility assignments in the intended actions phase. This generalization is not pursued here, because it is not needed for the kind of situations and issues considered in this section, particularly for most situations and issues involving emotional states, where changes in subjective probabilities appear to be a more reasonable choice.

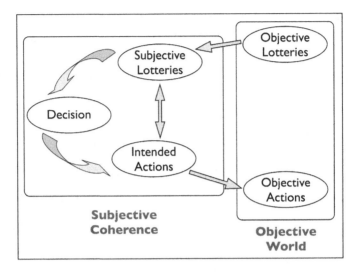

Fig. 5.2   Subjective Rationality

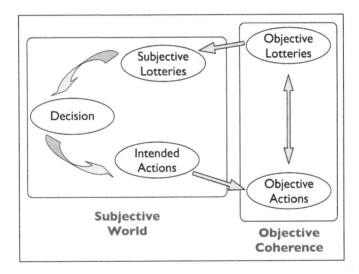

Fig. 5.3   Objective Rationality

tery influences together with influences determined by outside factors (e.g., the decision maker is under the influence of a particular drug) are combined into a single *decision state (for L)*— called a *state* for short—where for each lottery L, the decision state for L determines how the decision maker subjectively represents and assigns subjective probabilities to the events within L.

Definition 5.2. Let S denote the set of decision states. $\square$

Let s be an element of S and $L = (a_1, A_1; \ldots; a_i, A_i; \ldots; a_n, A_n)$ be an objective lottery that the decision maker is evaluating while in state s. Then, by definition,

- $L^s = (a_1^s, A_1^s; \ldots; a_i^s, A_i^s; \ldots; a_n^s, A_n^s)$ is the decision maker's cognitive representation of L when he is in state s.

L^s is called a *subjective lottery,* and all subjective lotteries have this form. Let $\mathcal{L}^S$ stand for the set of subjective lotteries. DSEU assumes the following invariances:

- *Invariance of cognitive representations of lotteries:* Each subjective lottery is formally a lottery with pure outcomes, and it has the same number of events as the objective lottery of which it is a representation.
- *Invariance of cognitive representations of outcomes:* For all s and t in S and all pure outcomes a in an objective lottery, $a^s = a^t$.
- *Invariance of disjointness across states:* Suppose events A and B occur in objective lotteries and s and t are in S. Then
$$A \cap B = \varnothing \quad \text{iff} \quad A^s \cap B^t = \varnothing. \tag{5.9}$$

Note that Equation 5.9 implies a strong relatedness across states of the cognitive representations of events from objective lotteries. DSEU assumes the following about the cognitive representations of events from objective lotteries:

- *Open set interpretation of events:* $\mathcal{T}$ is a topology with universal event X and with operation of pseudo complementation, $\dot{-}$. The cognitive representations of events from objective lotteries are in $\mathcal{T}$.
- *Coherence of cognitive representations of events:* For all events A from objective lotteries and all s and t in S, $\dot{-}\dot{-}A^s = \dot{-}\dot{-}A^t$.

DSEU assumes the following about the utility of subjective lotteries:

- *SEU utility for subjective lotteries:* As part of the intended actions, the utility of a subjective lottery $L^s = (a_1^s, A_1^s; \ldots; a_i^s, A_i^s; \ldots; a_n^s, A_n^s)$ is computed by
$$u(L^s) = \sum_{i=1}^{n} \frac{\mathsf{P}_s(A_i^s)}{\mathsf{P}_s(A_1^s) + \cdots + \mathsf{P}_s(A_n^s)} \cdot u(a_i), \tag{5.10}$$

where P_s is a function from $\{A_1, \ldots, A_n\}$ into $[0,1]$ such that $\sum_{i=1}^n \mathsf{P}_s(A_i) = 1$. Note that P_s in Equation 5.10 depends on the state s. The later assumption below of probabilistic coherence will remove this dependence. u does not depend on s, because of the assumption of invariance of cognitive representation of outcomes.

- *Subjective preference:* Suppose the decision maker is required to judge which of the two objective lotteries L and M he prefers. He does this while in state s. Thus he represents them cognitively as L^s and M^s. Later, in the intended action stage of decision process, he intends to indicate that he prefers the larger of $\{u(L^s), u(M^s)\}$, and, still later, he indicates this preference in the outside world by indicating L if $u(L^s)$ is larger, and indicating M if $u(M^s)$ is larger.

To make DESU close to SEU while allowing for states of the decision maker to influence his decision making, the following assumption of probabilistic coherence is added. It is use later to argue for the rationality of DSEU.

- *Probabilistic coherence:* The following two statements are assumed:
  (1) The events occurring in lotteries in L^S are elements of a boolean algebra of events with an L-probability function P.
  (2) For each event A^s occurring in some lottery L^s in L^S, $\mathsf{P}(A^s) = \mathsf{P}_s(A^s)$, where P_s is as defined in the above "SEU utility for subjective lotteries" (Equation 5.10).

The above assumptions constitute the DSEU model. It is a theory of *subjective rationality*. It is not a theory of *objective rationality,* because in the outside world the SEU model may not hold, for example, because of a change of state, the decision maker may strictly prefer an objective lottery L to an objective lottery M and other times strictly prefer M to L. Of course, if DSEU holds, and for all L in $\mathcal{L}^S$ and all s and t in S, $L^s = L^t$, then the SEU model would hold in the outside world as well. For such a situation, *objective rationality* is said to hold. SEU holding for objective lotteries is viewed as a form of *objective coherence.*

The following is an argument for the rationality of DSEU. Assume DSEU. By Theorem 4.16, there exists an extended probability function P' on the power-set Y of the domain of P that is an extension of P. Let L_Y be the set of all lotteries (with finitely many events) that have events from Y and pure outcomes from lotteries in L^S. Then all the subjective lotteries of DSEU are in L_Y. Define the weak ordering $\precsim_Y$ on L_Y and $\precsim_{sub}$ on L^S as follows: For all lotteries $L = (a_1, A_1; \ldots; a_i, A_i; \ldots; a_n, A_n)$ and

$M = (b_1, B_1; \ldots; b_j, B_j; \ldots; b_m, B_m)$, in L_Y, let

$$L \precsim_Y M \quad \text{iff}$$

$$\sum_{i=1}^{n} \frac{\mathsf{P}'(A_i)}{\mathsf{P}'(A_1) + \cdots + \mathsf{P}'(A_n)} \cdot u(a_i) \leq \sum_{i=j}^{m} \frac{\mathsf{P}'(B_j)}{\mathsf{P}'(B_1) + \cdots + \mathsf{P}'(B_m)} \cdot u(b_j),$$

and for all lotteries $H = (c_1, C_1; \ldots; c_p, C_p)$ and $K = (d_1, D_1; \ldots; d_q, D_q)$ in L^S, let

$$H \precsim_{sub} K \quad \text{iff}$$

$$\sum_{i=1}^{p} \frac{\mathsf{P}'(C_i)}{\mathsf{P}'(C_1) + \cdots + \mathsf{P}'(C_p)} \cdot u(c_i) \leq \sum_{j=1}^{q} \frac{\mathsf{P}'(D_j)}{\mathsf{P}'(D_1) + \cdots + \mathsf{P}'(D_q)} \cdot u(d_j).$$

Then $\precsim_Y$ satisfies the SEU. Consequently, DSEU is consistent with SEU in the following sense: Any irrationality directly occurring in DSEU's preference ordering $\precsim_{sub}$ is an irrationality occurring SEU's preference ordering $\precsim_Y$. In this sense, DSEU's preference ordering $\precsim_{sub}$ is "rational". This is consistent with how SEU is used to describe empirical, decision phenomena: *Rationality is said to hold if the decision maker's preference behavior does not reveal irrationality.*

DSEU generalizes SEU by allowing some objective events to have many subjective interpretations and thus to have many subjective probability assignments. This is done while having the subjective utility assignments to pure outcomes remain invariant with respect to probability assignments. In order to explain some kinds of violations of SEU, various researchers have modeled violations of SEU by having invariant probability assignments while emphasizing utility features of the outcome, often about whether an outcome is viewed positively (e.g., as a gain) or negatively (e.g., as a loss). This is illustrated in the following experiments of Tversky and Kahneman involving students from Stanford University and University of Vancouver. They were given one of the following two scenarios.

> **Problem 1.** Imagine that the U.S. is preparing for the outbreak of an unusual Asian disease, which is expected to kill 600 people. Two alternative programs to combat the disease have been proposed. Assume that the exact scientific estimate of the consequences of the programs are as follows: If Program A is adopted, 200 people will be saved. If Program B is adopted, there is $\frac{1}{3}$ probability that 600 people will be saved, and $\frac{2}{3}$ probability that no people will be saved. Which of the two programs would you favor?
> *152 subjects participated in Problem 1. 72% of them chose Program A and 28% of them chose Program B.*

Problem 2. The subjects given this scenario is given the same cover story of Scenario 1 but with the two alternative programs formulates as follows: If Program C is adopted 400 people will die. If Program D is adopted there is $\frac{1}{3}$ probability that nobody will die, and $\frac{2}{3}$ probability that 600 people will die. Which of the two programs would you favor?

155 subjects participated in Problem 2. 22% of them chose Program C and 78% of them chose Program D.

Tversky & Kahneman (1981) write the following about these experiments:

> The majority choice in problem 1 is risk averse: the prospect of certainly saving 200 lives is more attractive than a risky prospect of equal expected value, that is, a one-in-three chance of saving 600 lives. ... The majority choice in problem 2 is risk taking: the certain death of 400 people is less acceptable than the two-in-three chance that 600 will die. The preferences in problems 1 and 2 illustrate a common pattern: choices involving gains are often risk averse and choices involving losses are often risk taking. However, it is easy to see that the two problems are effectively identical. The only difference between them is that the outcomes are described in problem 1 by the number of lives saved and in problem 2 by the number of lives lost. The change is accompanied by a pronounced shift from risk aversion to risk taking. We have observed this reversal in several groups of respondents, including university faculty and physicians. *(p. 453)*

Various models have been put forth to describe how gains are treated differently the losses. These generally are "sign-dependent" models which employ different utility functions for objects of having positive value and thoses having negative value. An example of this is the following sign-dependent model of Luce (1991). (Luce derives this model from qualitative assumptions for the utility of a gamble.)

The following is the formula Luce derives for the sign-dependent case involving the utility of gambles with two outcomes:

$$u[(\$a, A; \$b, B)] = \begin{cases} u(\$a)w^+(A) + u(\$b)(1 - w^+(B)) & \text{if } \$a > \$0,\ \$b > \$0, \\ u(\$a)w^-(A) + u(\$b)(1 - w^-(B)) & \text{if } \$a < \$0,\ \$b < \$0, \\ u(\$a)w_1(A) + u(\$b)(1 - w_1(B)) & \text{if } \$a > \$0,\ \$b \leq \$0, \\ u(\$a)w_2(A) + u(\$b)(1 - w_2(B)) & \text{if } \$a \leq \$0,\ \$b > \$0. \end{cases}$$
$$(5.11)$$

The model described in Equation 5.11 is for descriptive modeling, and, for this reason, w^+, w^-, w_1, and w_2 are called "weighting functions," because they need not be additive over disjoint events.

For comparison purposes, consider the following is a DSEU model of a closely related situation: The participant has two states, one when $\$c \geq \$0$, yielding the subjective probability $P^+(C)$ for C, and one when $\$c < \$0$, yielding the subjective probability $P^-(C)$ for C. Then for each binary gamble $(a, A; b, B)$,

if $\$a \geq \$0$ and $\$b \geq \$0$, then

$$u[(a, A; b, B)] = \frac{P^+(A)u(a)}{P^+(A) + P^+(B)} + \frac{P^+(B)u(b)}{P^+(A) + P^+(B)} \,;$$

if $\$a < \$0$ and $\$b < \$0$, then

$$u[(a, A; b, B)] = \frac{P^-(A)u(a)}{P^-(A) + P^-(B)} + \frac{P^-(B)u(b)}{P^-(A) + P^-(B)} \,;$$

if $\$a \geq \$0$ and $\$b < \$0$, then

$$u[(a, A; b, B)] = \frac{P^+(A)u(a)}{P^+(A) + P^-(B)} + \frac{P^-(B)u(b)}{P^+(A) + P^-(B)} \,;$$

if $\$a < \$0$ and $\$b \geq \$0$, then

$$u[(a, A; b, B)] = \frac{P^-(A)u(a)}{P^-(A) + P^+(B)} + \frac{P^+(B)u(b)}{P^-(A) + P^+(B)} \,.$$

Narens (2011) comments,

> The DSEU and Luce's models are very similar. The principal difference is that DSEU requires much more additivity than Luce's model, because P is required to be a lattice probability function. This makes the DSEU model much more rational or normative-like than the Luce model, and thus a better model for matters such as public policy decisions, where loss aversion is included as a priority. On the other hand, for laboratory studies of gambling with relatively small sums of money, Luce's model is likely to describe data better than DSEU, because of well-known failures of probabilistic additivity in related studies.
>
> The above DSEU model, which has two probability functions, does not distinguish between some pairs of gambles having both positive and negative outcomes in the manner of the Luce model that has four probability functions. This distinction, if needed, can be incorporated into [an extension of DSEU having four states].

There are other models in the literature that stress the impact of utilities on subjective probability. For example, Lopes (1987) examines the impacts of hope and fear on decision making. Interpreted in terms of DSEU, Lopes modeling amounts to the following: A large gain puts the decision maker in a hope state, which increases the subjective probability associated with the gain. A large loss put the decision maker in a fear mode, which also increases the subjective probability associated with the loss. Chichilnisky (2009) develops a theory about the impact of catastrophes on decision making. Interpreted in terms of DSEU, a catastrophic outcome o induces a fear state in the decision maker, which increases his subjective probability of o occurring. Chichilnisky develops a novel probability theory to account for this kind of increase in subjective probability.

In summary, there are three obvious ways of incorporating states into the decision making process: (i) subjective probability functions that vary with state, (ii) utility functions that vary with state, and (iii) a combination of (i) and (ii). The literature has mainly focused on (ii), with a few cases where the focus is on (i). There has not been, to my knowledge, experimental paradigms and studies to distinguish among (i), (ii), and (iii). The DSEU developed in this section is a special kind of model for (i). It satisfies strong coherence conditions and is a well-defined, explicit model of bounded rationality.

Chapter 6

Orthomodular Modeling of Psychological Paradigms

6.1 Orthoprobability Theory

Following Boole (1854), Kolmogorov (1933), and von Mises (1928), probability theory has been traditionally formulated in terms of a boolean algebra of events. In order to match probabilities to observable quantum phenomena, Dirac (1930) and von Neumann (1932) instead employed closed subspaces of a Hilbert space as events in their formulations of probability theory for quantum phenomena. Birkhoff & von Neumann (1936) decided that orthocomplemented modular lattices provided an abstract algebraic description of the event space underlying quantum phenomena. They called their description *quantum logic*. But Husimi (1937) showed that this was wrong: While Birkhoff's and von Neumann's ideas were valid for analogs of finite dimensional Hilbert spaces, they were not general enough for infinite dimensional Hilbert space. For general Hilbert space, Husimi showed that the event lattice needed to satisfy the orthomodular law, and he suggested that the logical structure of event spaces described by Birkhoff's & von Neumann's "quantum logic" should instead be the theory of orthomodular lattices. His suggestion has been followed in the literature.

Geechie (1971) showed that some orthomodular lattices cannot have an orthoprobability defined on them. This suggests that orthocomplementation and the orthomodular law by themselves are not adequate for founding a scientifically useful theory of probability.

This chapter's approach is to assume an ortholattice *and* the existence of an orthoprobability function, and then derive, as a consequence, the orthomodular law. This method is useful for behavioral applications, because probabilities are usually collected as part of the data. As described later in the chapter, orthoprobability is a natural generalization for situations where there are gaps in the probabilistic descriptions of relationships across experiments.

Recall that an *orthoprobability function* $\mathbb{P}$ on an ortholattice

$$\mathfrak{L} = \langle L, \leq, \sqcup, \sqcap, ^\perp, 1, 0 \rangle$$

is a function into the extended real interval $^*[0,1]$ such that $\mathbb{P}(1) = 1$, $\mathbb{P}(0) = 0$, and for all a and b in L,

(*i*) if $a < b$ then $\mathbb{P}(a) < \mathbb{P}(b)$, and
(*ii*) if $b \leq a^\perp$ then $\mathbb{P}(a \sqcup b) = \mathbb{P}(a) + \mathbb{P}(b)$.

Most of this chapter concerns the special cases where $\mathbb{P}$ is taken to be a function into the real interval $[0,1]$. For these cases $\mathbb{P}$ is called a *traditional orthoprobability function*. Note that it follows that traditional orthoprobability functions are special cases of orthoprobability functions.

Recall that a lattice is *orthomodular* if and only if it is orthocomplemented and satisfies the orthomodular law,

for all a and b in L, if $a \leq b$ then $a \sqcup (a^\perp \sqcap b) = b$,

and that in Chapter 2 it was shown that for a lattice, distributivity implied modularity, and modularity implied orthomodularity.

Theorem 6.1. *Suppose* $\mathfrak{L} = \langle L, \leq, \sqcup, \sqcap, ^\perp, 1, 0 \rangle$ *is an orthocomplemented lattice and* $\mathbb{P}$ *is an orthoprobability function on it. Then* $\mathfrak{L}$ *is orthomodular.*

Proof. By Theorem 2.25 of Chapter 2 it needs only to be shown that $\mathfrak{L}$ has no O_6 subalgebra. Suppose $\mathfrak{L}$ had an O_6 subalgebra as displayed in Figure 6.1. A contradiction will be shown. Then $y^\perp < x^\perp$ and thus, by the definition of "orthoprobability" and the facts that

$$x \sqcup x^\perp = x \sqcup y^\perp = 1,$$

$$\mathbb{P}(x \sqcup x^\perp) = \mathbb{P}(x) + \mathbb{P}(x^\perp) = 1 \ \text{ and } \ \mathbb{P}(x \sqcup y^\perp) = \mathbb{P}(x) + \mathbb{P}(y^\perp) = 1 \quad (6.1)$$

and

$$\mathbb{P}(y^\perp) < \mathbb{P}(x^\perp). \qquad (6.2)$$

Because it is impossible for both Equations 6.1 and 6.2 to hold, a contradiction has been shown. $\square$

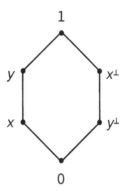

Fig. 6.1   Lattice O_6

6.2   Experimental Psychological Paradigms

6.2.1   *Formalism and lattice foundation*

This chapter considers two types of psychological paradigms: A between-subjects paradigm and an individual subject paradigm. In a between-subjects paradigm, experiments are conducted on different groups of subjects and conclusions are reached by comparing the results of a group with other groups. In an individual subject experiment, a subject makes choices in varying mental states (contexts) and conclusions are reached by comparing the results across contexts. Formally these two kinds of paradigms are structurally the same, with subjects in a between-subjects experiment corresponding to the mental states of the individual subject. For the most part, matters are formulated for between-subjects paradigms.

A *between-subjects experimental psychological paradigm,* or just *paradigm* for short, consists of the following:

- SUB = a nonempty, finite set of subjects.
- EXP = $\{E_i \,|\, i \in I\}$, an indexed set of experiments, where I is a nonempty finite set of indices.
- S_i = the set of subjects participating in experiment E_i.
- $O_i = \{o_{i,j} \,|\, j \in J(i)\}$ = the finite set of outcomes of experiment E_i, where for each i in I, $J(i)$ is a nonempty set of indices.
- $O = \bigcup_{i \in I} O_i$, the set of the paradigm's outcomes.

It is assumed that each O_i is nonempty and that each subject S_i chooses exactly one of its outcomes. For example, suppose $o_{i,j}$ is the action, "Pressed the button on the left in experiment E_i", and $o_{h,k}$ is the action, "Pressed the button on the left in experiment E_h", where $i \neq h$. Then even though $o_{i,j}$ and $o_{h,k}$ describe the action of pressing the left button, $o_{i,j} \neq o_{h,k}$, because they take place in different experiments. Thus it is assumed that

$$O_i \cap O_h = \varnothing \text{ for all } i \neq h, i \text{ and } h \text{ in } I.$$

Further, it is assumed that all the outcomes in O_i are distinct, that is,

$$o_{i,j} \neq o_{i,k} \text{ for all } j \text{ and } k \text{ in } J(i), j \neq k.$$

This section's formalism for psychological experimentation deviates a little from the usual ways they are described. In particular, subjects' choices made during experimentation are partially accounted for by the indexing of experiments. That is, what is in common use looked upon as an "experiment" is called here a "paradigm", which, by this chapter's formalism, is an indexed family of experiments. For example, suppose in an "experiment" each subject is put into one of two conditions: (a) Reward, or (b) No Reward. In the Reward condition he is given $100 and is asked to choose between alternatives C and D. In the No Reward condition his given $0 and is asked to choose among the alternatives D, F, and G. For the purposes of our formalism, this "experiment" is viewed as two experiments: E_a, the experiment consisting of those subjects who received $100 and has as its outcomes alternatives C_a and D_a, and experiment E_b consisting of those subjects who received $0 and has as its outcomes alternatives D_b, F_b, and G_b. Note that the common alternative D in the "experiment" is treated by our formalism as two different outcomes, D_a and D_b, each belonging to a different experiment. This is done to emphasize that subjects who received $100 were in a systematically different context that may have influenced their decision in choosing D_a than those who received $0 when deciding whether or not to choose alternative D_b. If it is a hypothesis of the paradigm that the difference in context does not matter—effectively making D_a and D_b the same choice—then the subscripting is maintained but an explicit hypothesis is added to the paradigm's theory that is stated in counterfactual terms, for example, "Each subject in experiment E_a who chooses D_a would have chosen D_b if he were put in experiment E_b, and similarly, each subject in experiment E_b who chooses D_b would have chosen D_a if he were put in experiment E_a." This strategy allows the chapter's formalism to include an effective equivalent to "different experiments having outcomes in common" while maintaining indexing. This also provides an equivalent to the idea that "The choice of D did not depend on the experimental context."

The formalism for a between-subjects paradigm has two parts: (i) The *data*, which consists of which subject participated in which experiment and which outcome she chose, and (ii) the *theory*, which consists of counterfactual statements relating subjects' behavior across the paradigm's experiments. The approach to "common outcomes" described previously is part of the paradigm's theory.

The *paradigm's data* is about what each subject does in each of the paradigm's experiments. It specifies the following:

- S_i, the set of subjects participating in E_i for each i in I.
- $< o_i >$, the set of subjects who chose outcome o_i in experiment E_i. It is assumed that $o_i \neq o_j$ for all $i \neq j$.
- For A in the power-set of O_i, $\wp(O_i)$, $< A >$ is the set of paradigm's subjects in experiment E_i who chose some element of A. This includes as special cases, $< O_i > = S_i$, the set of subjects who chose some element of O_i.

A major part of scientific thinking about paradigms is concern with the linkages of behaviors across paradigm experiments. In between-subjects experiments, such linkages are not directly observable: They result from a combination of what is observed within an experiment and a theory about how these observations are linked across experiments. The following concepts are used in formulating the paradigm's theory:

- Let o be an outcome in O_i and s be a subject in SUB. Then s is said to have *actually chosen* o if and only if s participated in experiment E_i and chose o. s is said to have *counterfactually chosen* o if and only if s did not participate in E_i but would have chosen o if she would have had participated in E_i. *It is assume that a counterfactual choice exists for each subject in each paradigm experiment that she did not participate in. Sometimes a subject's counterfactual choice can be deduced from the paradigm's theory and data. Other times it is unknown.*
- Let s be a subject, A be an event in $\wp(O)$, and o be an outcome O. Then s is said to *behaviorally select* o *as an element of A* if and only if the paradigm's theory and data implies one of the following two statements:
  - (i) $o \in A$ and s actually chose o.
  - (ii) $o \in A$ and s counterfactually chose o.

- Let s be a subject and A be an element of $\wp(O)$. Then s is said to *behaviorally select* A if and only if

  (i) for each o in A, s behaviorally selects o as an element of A, and

  (ii) for each o in $-A$, where $-$ is the set-theoretic complement of A with respect to O, s behaviorally selects o as an element of $-A$.

- Let A be in $\wp(O)$. Then A is said to be a *determinable event* if and only if for each paradigm subject s, s behaviorally selects A or s behaviorally selects $-A$.

- The notation $\ll A \gg$ stands for the set of paradigm subjects who for the determinable event A behaviorally selected some element of A.

Let A be a determinable event. Note that it follows from the definition of $\ll A \gg$ that the paradigm's theory and data exactly specify which of paradigm's subjects are in $\ll A \gg$.

- For an arbitrary finite set T, let $|T|$ = the *size of T* = the number of elements of T.

Because it is known for each subject s whether or not $s \in \ll A \gg$, it follows that the size of $\ll A \gg$, $|\ll A \gg|$, is exactly known. A probability function $\mathbb{P}$ is defined on the set of determinable events as follows: For each determinable event A,

$$\mathbb{P}(A) = \frac{|\ll A \gg|}{|O|}.$$

$\mathbb{P}$ is called *the paradigm's determinable probability function*. It assigns to each determinable event a number that is exactly determined by the paradigm's data and theory.

Because in experiment E_i each subject in S_i has to choose exactly one element in O_i, it follows that for A and B in $\wp(O_i)$,

$$A = B \quad \text{iff} \quad \ll A \gg = \ll B \gg .$$

But a similar result need not hold for A and B in $\wp(O)$. For example, for $i \neq j$, $O_i \neq O_j$, but $\mathsf{SUB} = \ll O_i \gg = \ll O_j \gg$. For scientific analyses of outcome events C and D in $\wp(O)$, it is natural to view C and D describing the same (actually or counterfactually defined) situation if and only if

$$\ll C \gg = \ll D \gg .$$

This suggests defining the equivalence relation $\equiv$ on determinable events as follows: For all determinable events F and G,

$$F \equiv G \quad \text{iff} \quad \ll F \gg = \ll G \gg .$$

When thinking about paradigm experiments, $\equiv$-equivalence classes are the natural objects for probabilistic assignment. It is easy to verify each $\equiv$-equivalence class $\mathcal{F}$ has a $\subseteq$-largest element, $\bigcup \mathcal{F}$. These largest elements are called *propositions*. It turns out that it is more convenient to assign probabilities to propositions instead of their $\equiv$-equivalent classes.

Propositions are particular determinable events. In general, the set of determinable events do not have an appropriate algebraic structure to be the domain of a mathematically rich probability function. However, the set of propositions has such an algebraic structure. To demonstrate this, the following definitions and notation are needed:

- Π = the set of propositions.
- $\pi(A)$, for each determinable event A, stands for the unique proposition in Π such that $\pi(A) = \ll A \gg$.
- $\sigma(\ll A \gg)$, for each determinable event A, is defined by,

$$\sigma(\ll A \gg) = \pi(A).$$

Thus, for each determinable event A, $\sigma(\ll A \gg) = \ll A \gg$. Note that if A is a proposition, then $\pi(A) = A$, and thus

$$\sigma(\ll A \gg) = A. \tag{6.3}$$

The restriction of $\mathbb{P}$ to Π is called *the paradigm's propositional probability function*. For convenience, $\mathbb{P}$ will also be used to denote the paradigm's propositional probability function in addition to its use as the paradigm's determinable probability function. It will be easy to determine by context which interpretation of $\mathbb{P}$ is intended.

The following numbered paragraphs demonstrate that the event space $\langle \Pi, \subseteq \rangle$ provides the basis for the paradigm's propositional probability function to have a rich and interesting probability theory.

(1) *O and $\varnothing$ are in* Π. This is immediate from the above definitions.

$\square$

(2) *Suppose that A is an arbitrary element of* Π. *Then A the set-theoretic complement of A (with respect to O), $-A$, is in* Π *and*

$$\ll -A \gg = - \ll A \gg. \tag{6.4}$$

Proof. It follows from the definition of "proposition" that A is determinable, and it follows from the definitions of "determinable" and "behaviorally selected" that $-A$ is determinable.

It will first be shown that

$$\ll A \gg \cap \ll -A \gg \; = \; \varnothing . \tag{6.5}$$

Suppose $s \in (\ll A \gg \cap \ll -A \gg)$. A contradiction will be shown. Because $s \in \ll -A \gg$, let o in $-A$ be such that s behaviorally selects o. Because A is a determinable event, each outcome in A that is behaviorally selected by each subject in $\ll A \gg$ is in A. Therefore, because by hypothesis s is also in $\ll A \gg$, it follows that o is in A, which is impossible since $o \in (-A)$.

Because A is a determinable event, it is immediate from the definitions of "determinable event" and "behaviorally selects" that

$$\ll A \gg \cup \ll -A \gg \; = \; \mathsf{SUB} . \tag{6.6}$$

Equations 6.5 and 6.6 show that for each proposition A, $\ll A \gg$ and $\ll -A \gg$ are set-theoretic complements (with respect to SUB) of each other, and therefore,

$$\ll -A \gg \; = \; - \ll A \gg . \tag{6.7}$$

Thus to complete the proof of (2), it needs to only be shown that $-A$ is a proposition. This will be done by contradiction. Suppose $-A$ were not a proposition. Let B be a proposition such that $-A \subset B$ and $\ll -A \gg \; = \; \ll B \gg$. Then let o be an element of $B - (-A) = B \cap A$, and, because B is a proposition, let t be an element of $\ll B \gg$ such that t behaviorally selects o. Because, by hypothesis, $\ll -A \gg \; = \; \ll B \gg$, it follows that

$$t \in \ll -A \gg . \tag{6.8}$$

Because $o \in (A \cap B)$,

$$o \in A .$$

Because A is a proposition and $o \in A$ and, by hypothesis, t behaviorally selects o, it follows that

$$t \in \ll A \gg . \tag{6.9}$$

Thus by Equations 6.8 and 6.9,

$$t \in (\ll -A \gg \cap \ll A \gg) ,$$

contradicting Equation 6.5. □

(3) $\langle \Pi, \subseteq, O, \varnothing \rangle$ *is a lattice.* O and $\varnothing$ are in Π by Paragraph (1). It is immediate that they are the unit and zero elements of $\langle \Pi, \subseteq \rangle$. Suppose A and B are arbitrary elements of Π. Then from the paradigm's theory

and data it is known for each subject s whether or not s is in $\ll A \gg$ and whether or not s is in $\ll B \gg$. Thus it is known for each subject s whether or not s is in $\ll A \gg \cup \ll B \gg$ and whether or not s is in $\ll A \gg \cap \ll B \gg$. Let

$$A \uplus B = \sigma(\ll A \gg \cup \ll B \gg) \quad \text{and} \quad A \sqcap B = \sigma(\ll A \gg \cap \ll B \gg).$$
(6.10)

Then, by the definition of σ, $A \uplus B$ and $A \sqcap B$ are in Π. The following shows that $A \uplus B$ is the $\subseteq$-l.u.b in Π of A and B. Suppose C in Π is such that

$$\ll A \gg \subseteq \ll C \gg \quad \text{and} \quad \ll B \gg \subseteq \ll C \gg.$$

Then

$$\ll A \gg \cup \ll B \gg \subseteq \ll C \gg,$$

and therefore,

$$A \uplus B = \sigma(\ll A \gg \cup \ll B \gg) \subseteq \sigma(\ll C \gg) = C.$$

A similar argument shows that $A \sqcap B$ is the $\subseteq$-g.l.b in Π of A and B. □

(4) *The following two statements hold:*

(i) $\langle \Pi, \subseteq, \uplus, \sqcap, -, O, \varnothing \rangle$ *is a complemented lattice, where* $-$ *is set-theoretic complementation.*

(ii) *Let A and B be arbitrary elements of Π such that $B \subseteq -A$. Then*

$$\ll A \uplus B \gg = \ll A \gg \cup \ll B \gg.$$
(6.11)

(*i*). By paragraph (3), $\langle \Pi, \subseteq, \uplus, \sqcap, O, \varnothing \rangle$ is a lattice. Using results from paragraph (2), it is shown that $-$, the operation of set-theoretic complementation, is a complementation operation of $\langle \Pi, \subseteq, \uplus, \sqcap, O, \varnothing \rangle$:

$$A \uplus (-A) = \sigma(\ll A \cup (-A) \gg) = \sigma(\mathsf{SUB}) = O$$

and

$$A \sqcap (-A) = \sigma(\ll A \cap (-A) \gg) = \sigma(\varnothing) = \varnothing.$$

(*ii*). It is immediate from the definitions of Π, $\ll \gg$, and $\uplus$ that

$$\ll A \gg \cup \ll B \gg \subseteq \ll A \uplus B \gg.$$

Thus, to show Equation 6.11, it needs only be shown that

$$\ll A \uplus B \gg \subseteq \ll A \gg \cup \ll B \gg.$$
(6.12)

Assume s is an arbitrary subject in $\ll A \uplus B \gg$. Because $-$ is the operation of set-theoretic complementation and $A \in \Pi$, it follows that either $s \in \ll A \gg$ or $s \in - \ll A \gg$. If $s \in \ll A \gg$, then $s \in (\ll A \gg \cup \ll B \gg)$. If $s \in - \ll A \gg$, then s did not behaviorally select any element in A. Because $s \in \ll A \cup B \gg$, s must have then behaviorally chosen some element of B, that is, $s \in \ll B \gg$, and thus $s \in (\ll A \gg \cup \ll B \gg)$. This shows Equation 6.12. $\square$

(5) $\mathfrak{P} = \langle \Pi, \subseteq, \uplus, \Cap, -, O, \varnothing \rangle$ *is an ortholattice, where* $-$ *is set-theoretic complementation.* To show DeMorgan's Laws, suppose A and B are in Π. Then, by Paragraph (3), $A \uplus B$ and $A \Cap B$ are in Π, and, by Equation 6.4 of Paragraph 2, $-(A \uplus B)$ and $-(A \Cap B)$ in Π. Note that because $-(A \uplus B)$ is in Π, it follows by Equations 6.3 and 6.7 that

$$-(A \uplus B) = \sigma(\ll -(A \uplus B) \gg) = \sigma[-(\ll (A \uplus B) \gg)]. \qquad (6.13)$$

By a similar argument,

$$-(A \Cap B) = \sigma(- \ll A \cap B \gg). \qquad (6.14)$$

By Equation 6.4 of Paragraph 2,

$$- \ll A \cup B \gg = \ll -(A \cup B) \gg \quad \text{and} - \ll A \cap B \gg = \ll -(A \cap B) \gg$$

and

$$- \ll A \gg = \ll -A \gg \quad \text{and} - \ll B \gg = \ll -B \gg.$$

Thus by DeMorgan's Law applied to the boolean algebra of events $\langle \wp(O), \cup, \cap, O, \varnothing \rangle$,

$$- \ll A \cup B \gg = \ll -(A \cup B) \gg = \ll (-A) \cap -(B) \gg$$

and

$$- \ll A \cap B \gg = \ll -(A \cap B) \gg = \ll (-A) \cup -(B) \gg,$$

and by Equation 6.13,

$$-(A \uplus B) = \sigma(- \ll A \cup B \gg) = \sigma(\ll (-A) \cap -(B) \gg) = (-A) \Cap (-B),$$

and by Equation 6.14,

$$-(A \Cap B) = \sigma(- \ll A \cap B \gg) = \sigma(\ll (-A) \cup -(B) \gg) = (-A) \uplus (-B). \quad \square$$

(6) $\mathbb{P}$ *is an orthoprobability function on* $\mathfrak{P} = \langle \Pi, \subseteq, \uplus, \Cap, -, O, \varnothing \rangle$. By Paragraph 5, $\mathfrak{P}$ is an ortholattice. Let A and B be arbitrary elements of Π such that $B \subseteq -A$. Because $\ll B \gg \subseteq - \ll A \gg$, it follows that

$\ll A \gg \cap \ll B \gg = \varnothing$. Therefore, it follows from Equation 6.11 and the definition of $\mathbb{P}$ that

$$\mathbb{P}(A \uplus B) = \frac{|\ll A \uplus B \gg|}{|O|} = \frac{|\ll A \gg \cup \ll B \gg|}{|O|}$$

$$= \frac{|\ll A \gg|}{|O|} + \frac{|\ll B \gg|}{|O|} = \mathbb{P}(A) + \mathbb{P}(B),$$

and thus that $\mathbb{P}$ is an orthoprobability function on $\mathfrak{P}$. $\square$

$\mathfrak{P}$ in paragraph (6) is called *the paradigm's propositional lattice*.

Lattices that are isomorphic to a paradigm propositional lattice routinely appear in quantum mechanics. For example, consider the lattice in Figure 6.2, which is from Example III of Foulis & Randall (1972). Foulis & Randall make the following comment (with this section's terminology used in place of their's) about this example:[1]

> Suppose we have a device that, from time to time, emits a particle and projects it along a linear scale. We consider two experiments, E_1 and E_2 defined as follows: In E_1 we look to see if there is a particle present. If there is not, we record the outcome of E_1 as the symbol n. If there is, we measure its position coordinate x. If $x \geq 1$, we record the outcome of E_1 as the symbol a, while if $x < 1$, we record the outcome of E_1 as the symbol b. Thus, the set of outcomes of E_1 is $O_1 = \{n, a, b\}$. In E_2 we look to see if there is a particle present. If there is not, we record the outcome of E_2 as the symbol n. If there is, we measure the x-component p_x of its momentum, recording the symbol c as the outcome of E_2 if $p_x \geq 1$ and the symbol d as the outcome of E_2 if $p_x < 1$. Thus, the set of outcomes for E_2 is $O_2 = \{n, c, d\}$. (The reason for our identification of the outcome n of E_1 with the outcome n of E_2 should be clear to the reader.) *(p. 1674)*

It is easy to imagine situations in behavioral science that produce lattices isomorphic to the one presented in Figure 6.2: Consider the situation where there are two committees, 1 and 2, with no member in common that are assigned to choose among the three candidates x, y, and z to lead a company. x and y have similar visions for the company and z has a radically different vision. Assume the choice is done by a vote, with each committee member of each committee providing a single vote for her preferred candidate. Then, using this section's terminology, this is modeled by a paradigm

---

[1] Foulis & Randall (1972) developed a different kind of formalism for generalizing aspects of quantum mechanics that ultimately arrived at concepts similar to the ones presented in this section, and Wright (1990) extended their formalism to include a random generation of outcomes through a classical urn model.

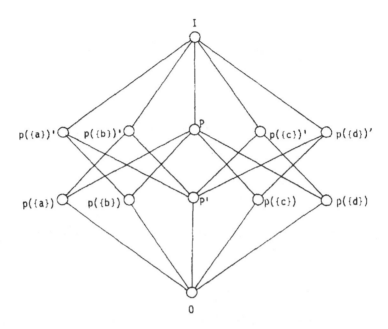

Fig. 6.2 The lattice of propositions for Foulis & Randall example. In this figure, $I =$ $O = \{a, b, n, c, d\}$, $0 = \varnothing$, $'$ is the lattice complementation operation, for each e in O, $p(e) = \pi(\{e\})$, $P' = p(n)$, and $P = \{a, b, c, d\} = p(n)'$. From the *Journal of Mathematical Physics*, Vol. 13, No. 11, November 1972, page 1674.

with its set of "subjects", SUB, consisting the union of the members of the two committees, its "experiments" E_i, $i = 1, 2$, having as "subjects" in E_i the members of committee i, and E_i "outcomes" being the candidates voted on in Committee i, denoted here by x_i, y_i, z_i, to emphasize that x_i is candidate x on a ballot for Committee i, etc. Assume the paradigm's "theoretical assumptions" consist of the following: For each s in SUB, if s voted for z_1 then she would have counterfactually voted for z_2, and if she voted for z_2 then she would have counterfactually voted for z_1. Then this paradigm produces a lattice of propositions isomorphic to the lattice in Figure 6.2.

6.2.2 *Between-subject paradigms and modeling the individual subject*

In a *between-subjects paradigm*, each subject participates in exactly one experiment. The previously discussed two committee voting paradigm is an example. A between-subject paradigm's yields a sparse data set. Such spareness is a major contributing factor to the non-boolean character of the paradigm's probabilistic data.

In a *within-subject paradigm*, each subject participates in each experiment. In this type paradigm, the data set is full in the sense that for each subject s and each outcome o, it is in the data whether or not s chose o. For full data sets, the lattice of propositions becomes a boolean algebra of events. In psychological experiments, it is often the case that the choice of an outcome in one experiment influences the choice of an outcome in another, making the analysis and interpretation of paradigm's data difficult. Between-subjects paradigms are used by psychologists to avoid this issue.

A lattice foundation can be developed for an individual subject that has a mathematical structure that is very similar to a between-subject paradigm. For such a *individual subject paradigm*, a set of states STS is used in place of the set of subjects SUB. It is assumed that the decision maker can assume each state s in STS. Let $\{s_j \mid j \in J\}$ be an indexing of STS. Contexts or situations C_i that the subject could find herself in are used in place of experiments E_i of a between-subject paradigm. Let I be an indexing set of contexts. Each context C_i determines a set of outcomes O_i from which the individual subject makes a choice while in context i. Her choice depends not only on the choice set C_i, but also on the state s_j that she is in. The combination of C_i and s_j determine her choice $o_{i,j}$ from O_i. Let $J(i)$ be the set of states the subject can assume while in context C_i.

$o_{i,j}$ is assumed to be probabilistic in the sense that there is some probability that she will choose $o_{i,j}$ with probability $p_{i,j}$ when in context C_i. Because of this, $1 = \sum_{j \in J(i)} p_{i,j}$. Analogous to a between-subjects design, it is assume that the subject will make exactly one choice when in context C_i and state s_j.

It is important to keep in mind that the individual subject's choice behavior is different from subjects participating in a between-subjects experiment in that her choice is modeled probabilistically instead of deterministically. Also note that from a theoretical perspective it still makes sense to consider what her choice behaviors would have been in other contexts. The principal difference between individual-subject and between-subjects experiments is that the probability distribution on outcomes across contexts is theoretical and non-observable for the individual subject, whereas in a between-subjects paradigm the probability distribution on outcomes across experiments is empirical and observable. Outside of this, the two kinds of paradigms have identical mathematical structures. Most of the presentation in this chapter is made for between-subjects experiments, because they can be presented in a more concrete manner.

6.3   The Disjunction Effect

The *sure thing principle* is a hallowed principle of economic decision making. It is used here to provide a simple illustration of how the logical structure of an experimental psychological paradigm can be construed to be an orthomodular lattice.

Savage (1954) provides the following illustration of the sure thing principle.

> A businessman contemplates buying a certain piece of property. He considers the outcome of the next presidential election relevant to the attractiveness of the purchase. So, to clarify the matter for himself, he asks whether he would buy if he knew that the Republican candidate were going to win, and decides that he would do so. Similarly, he considers whether he would buy if he knew that the Democratic candidate were going to win, and again finds that he would do so. Seeing that he would buy in either event, he decides that he should buy, even though he does not know which event obtains .... It is all too seldom that a decision can be arrived at on the basis of the principle used by this businessman, but, except possibly for the assumption of simple ordering. I know of no other extralogical principle governing decision that finds such ready acceptance. (Savage, 1954, p. 21)

Savage used a version of the sure thing principle as a key axiom in his seminal theory of subjective expected utility. Tversky & Shafir (1992) designed various tests of the principle. They write the following about Savage's quotation above and its empirical validity:

> Savage went on to define this principle formally: If x is preferred to y knowing that event A obtained, and if x is preferred to y knowing that A did not obtain, then x should be preferred to y even when it is not known whether A obtained. This rule, which Savage called the *sure-thing principle* (henceforth STP), has a great deal of both normative and descriptive appeal. Nevertheless, this principle does not always hold, especially when the decision maker has different reasons for making the same decision in different states of the world .... (p. 305)

To test the sure thing principle, Tversky and Safir (1992) developed experimental paradigms. Using ideas from prospect theory (Kahneman and Tversky, 1979), they designed the paradigms so that they would likely violate the sure thing principle. Three of their paradigms are described as follows:

First Paradigm

The first paradigm has three experiments, a Won Experiment, a Loss Experiment, and a Disjunctive Experiment. Ninety-eight subjects took part in the Won Experiment. A week later they took part in the Lost Experiment, and 10 days after that they took part in the Disjunctive Experiment.

E_1: Won Experiment Imagine that you have just played a game of chance that gave you a 50% chance to win $200 and a 50% chance to lose $100. The coin was tossed and you have won $200.

You are now offered a second identical gamble: 50% chance to win $200 and 50% chance to lose $100. Would you:

    x. Accept the second gamble. 69%
    y. Reject the second gamble. 31%

E_2: Loss Experiment Identical to the Won Experiment except instead of being told that they won $200 when the coin was tossed, they are told, "The coin was tossed and you have lost $100." Would you:

    x. Accept the second gamble. 59%
    y. Reject the second gamble. 41%

E_3: Disjunctive Experiment Imagine that you have just played a game of chance that gave you a 50% chance to win $200 and a 50% chance to lose $100. Imagine that the coin has already been tossed, but that you will not know whether you have won $200 or lost $100 until you make your decision concerning a second, identical gamble:

    50% chance to win $200,
    50% chance to lose $100.

Would you:

    x. Accept the second gamble. 36%
    y. Reject the second gamble. 64%

Tversky and Saffir write the following about the first paradigm:

> These problems were embedded among several similar problems so that the logical relation among the three versions would not be transparent.

Subjects were instructed to treat each decision separately. The data show that a majority of subjects accepted the second gamble after having won the first gamble, and a majority accepted the second gamble after having lost the first gamble. Most subjects, however, rejected the second gamble when the outcome of the first was not known. This pattern of preference clearly violates Savage's STP; we call it the *disjunction effect*. (p. 306)

Second Paradigm Tversky and Saffir write the following about their second paradigm for investigating the disjunction effect:

To estimate the reliability of choice, we presented the basic gamble (accept or reject an even chance to win $200 or lose $100) to a comparable group of subjects twice, several weeks apart. Only19% of the subjects (N = 95) made different decisions on the two occasions. The 65% rejection rate in the accept/accept cell, therefore, cannot be attributed to unreliability. (p. 307)

Third Paradigm Tversky and Saffir write the following about their third paradigm for investigating the disjunction effect:

We have replicated the disjunction effect in a between-subjects design. Three different groups of 71 subjects were presented with the Won version, the Lost version, and the disjunction version. As in the within-subjects design, a majority (69%) accepted the gamble in the Won condition, and a majority (57%) accepted the gamble in the Lost condition, but only 38% accepted the gamble in the disjunctive condition. The finding that the distribution of choices was nearly identical in the within-subjects and the between-subjects designs indicates that the respondents in the former study evaluated each version independently, with no detectable effects of one version on the other. An additional group (N = 75) was asked whether they would accept the gamble (even chance to win $200 or lose $100) when there had been no previous play. Only 33% of the subjects accepted the gamble in this condition. A similar rate of acceptance for this gamble (30%, N = 230) was reported previously (Tversky & Bar-Hillel, 1983). (p. 307)

Tversky & Saffir (1992) explain the empirical results of the above three paradigms using concepts from prospect theory (Kahneman & Tversky, 1979). Their explanation essentially consists of the following:

- In the Won Experiment, the decision maker has already won $200. Thus even a loss on the second gamble would leave him $100 ahead. This makes playing the gamble very attractive.

- In the Loss Experiment, the decision maker is down $100, "so playing the second gamble offers a chance to 'get out of the red,' which many people find more attractive than accepting a sure loss of $100".
- In the disjunctive condition, the decision maker does not know if he decides to play the second gamble which she will receive: a certain gain or a certain loss. Tversky & Saffir write, "The uncertainty concerning the first gamble, we suggest, makes it harder to contemplate the implication of each outcome. For example, the decision maker may be sure that he wishes to play the second gamble if he wins the first, but he may be unsure about his preference if he loses the first gamble. Only by focusing exclusively on the latter possibility does the decision maker realize that he wishes to play the second gamble in this case as well." They propose that this is not what people do. They write, "Instead, not knowing whether they have won or lost the first gamble, people segregate the second gamble and evaluate it from their current position. This assumption is supported by the observation that the percentage of subjects who accepted the gamble in the disjunctive version (36%) was similar to that observed when no prior gamble was mentioned (33%). Thus, the second gamble in the disjunctive condition is evaluated as an equal chance to win $200 or lose $100. Because of loss aversion, this gamble is not acceptable."

There is a large experimental literature showing the disjunction effect and failures of the sure thing principle. These failures have been studied and replicated many times by a variety of investigators. One of the failures involves the prisoner's dilemma. Tversky and Saffir write,

> ... we described the disjunction effect in a one-shot prisoner's dilemma game, played on a computer for real pay-offs. Subjects (N = 80) played a series of prisoner's dilemma games, without feedback, each against a different unknown opponent supposedly selected at random from among the participants. In this setup, the rate of cooperation was 3% when subjects knew that the opponent had defected, and 16% when they knew that the opponent had cooperated. However, when subjects did not know whether their opponent cooperated or defected (as is normally the case in this game), the rate of cooperation rose to 37%! *(p. 309)*

Prospect theory is a quantitative model of decision making that Tversky and Saffir used to model and explain their data involving gambles. In prospect theory, gains are modeled in a quantitatively different manner than losses. This allowed Tversky and Saffir to account for the failure of the sure thing principle in terms of the decision maker's utility function. However, their prospect theory modeling did not provide an account of why the decision maker in the disjunctive condition did not pursue an evaluation of the second gamble if he were to lose the first one.

Pothos and Busemeyer (2009) developed a radically different approach to provide an account of this and model the data. They viewed traditional probability theory as being too restrictive for modeling the psychology involved in the disjunctive effect. In its place, they employed the formalism of probability theory from quantum mechanics. They write,

> The main problems in developing a convincing cognitive quantum probability model are to determine an appropriate Hilbert space and Hamiltonian. We attempted to present a satisfactory prescriptive approach to deal with these problems and so encourage the development of other quantum probability models in cognitive science. For example, the Hamiltonian is derived directly from the parameters of the problem (e.g. the pay-offs associated with different actions) and known general principles of cognition (e.g. reducing cognitive dissonance). Importantly, our model works: it is able to account for violations of the sure thing principle in ... the two-stage gambling task and leads to close fits to empirical data.

It is beyond this book's scope to go into the details of Pothos and Busemeyer's theory, but it is worthwhile to note that they based the psychological part of their theory on well-established parts of cognitive psychology known as *cognitive dissonance theory* and *interference theory*, and they based the quantitative part of their theory on quantum probability theory. In contrast, Tversky and Saffir employed Kahneman and Tversky's theory of heuristics and biases (in the guise of prospect theory) and used traditional probability theory to account for the data.

This section's modeling of psychological paradigms also allows for violations of the sure thing principle. Like Pothos and Busemeyer, it employs a non-boolean event space—but one that differs from quantum probability theory with a Hilbert space and Hamiltonian. More generally, this section's probability modeling also differs from probability theories based on events from a Hilbert space.

Consider the sure thing situation previously described by Savage. This gives rise to a psychological paradigm involving three experiments:

- Experiment 1, where each subject "asks whether he would buy if he knew that the Republican candidate were to going win, and decides that he would do so."

- Experiment 2, where each subject "considers whether he would buy if he knew that the Democratic candidate were going to win, and again finds that he would do so."
- Experiment 3, where each subject "seeing that he would buy in either event, he decides that he should buy, even though he does not know which event obtains."
- Let R stand for the event, "a Republican is elected," $-R$ for the event, "a Democrat is elected," and for $i = 1, \ldots, 3$, B_i for the event "buy the property" in Experiment i.
- For $i = 1, 2$, let $\mathfrak{X}_i$ be the 4-element boolean algebra of events generated by R and B_i, and let $\mathfrak{X}_3$ be the 16-element boolean algebra of events generated by R, B_1, B_2, and B_3.

For this paradigm, I assume that Savage's sure thing principle consists of the following theoretical assumptions, where $\leftrightarrow$ stands for "if and only if":

$$B_1 \leftrightarrow (R \text{ and } B_3) \quad \text{and} \quad B_2 \leftrightarrow [(-R) \text{ and } B_3] \qquad (6.15)$$

and

$$-B_1 \leftrightarrow [R \text{ and } (-B_3)] \quad \text{and} \quad -B_2 \leftrightarrow [(-R) \text{ and } (-B_3)]. \qquad (6.16)$$

In paradigm propositional space without the assumption of the sure thing principle,

$$\pi(B_3) \cap \pi(B_1) = \varnothing \quad \text{and} \quad \pi(B_3) \cap \pi(B_2) = \varnothing$$

and

$$\pi(-B_3) \cap \pi(-B_1) = \varnothing \quad \text{and} \quad \pi(-B_3) \cap \pi(-B_2) = \varnothing.$$

Assuming the sure thing principle this changes to,

$$\pi(B_3) \subseteq \pi(B_1) \quad \text{and} \quad \pi(B_3) \subseteq \pi(B_2)$$

and

$$\pi(-B_3) \subseteq \pi(-B_1) \quad \& \quad \pi(-B_3) \subseteq \pi(-B_2).$$

The following modification of the paradigm accounts for the systematic violation of the sure thing principle:

Anti-Sure Thing Principle: $B_1 \leftrightarrow [R \cap (-B_3)]$, $B_2 \leftrightarrow [(-R) \cap (-B_3)]$, $-B_1 \leftrightarrow [R \cap B_3]$, and $-B_2 \leftrightarrow [(-R) \cap B_3]$.

The anti-sure thing principle produces a paradigm lattice of propositions that is isomorphic to the one produced by the sure thing principle. Thus it is not the logical structure of propositions (e.g., the failure of distributivity) that is responsible for the disjunctive effect. According to the orthomodular modeling presented here, the environment in Experiment 3 causes the subject to have a different psychological interpretation of "would buy" than in Experiments 1 and 2. This results in the same objectively defined situation having different subjective interpretations. It is the relationship among these different interpretations that produce the disjunctive effect.

In Chapter 5, multiple interpretations were also used to explain decision phenomena. There, the multiple interpretations resulted from states being added or subtracted from an event through the availability heuristic. In this section's orthomodular modeling, it is done in a different manner. For example, in the modeling involving the sure thing principle, the subjective interpretation of B_1 in Experiment 1 causes B_1 to be changed into R and B_3 in Experiment 3. One interpretation of this change is that R and B_3 can have a different subjective quality—for example, a quality corresponding to a mixture of uncertainty and riskiness—than B_3. This quality results from the uncertainty of not knowing which candidate will be elected. One goal of this section is to provide a framework for describing and potentially modeling this and related qualities in a rigorous, systematic manner. It is beyond the framework's scope to describe under which conditions such qualities will have a particular definitive impact on a subject's decision.

6.4   Dynamic Version of the Allais Paradox

A well-known and studied failure of subjective expected utility (SEU) is the Allais paradox (Allais, 1953). In the static version, in Experiment 1 subjects are asked to choose whether they prefer to play gamble (a) or gamble (b) (where $\$1M$ is $\$1,000,000$):

(a) Receiving $\$1M$ for certain.
(b) 10% chance of $\$5M$, 89% chance of $\$1M$, 1% chance of $\$0$.

In Experiment 2 subjects are asked to choose whether they prefer to play gamble c or gamble d:

(c) 10% chance of $\$5M$, 90% chance of $\$0$.
(d) 11% chance of $\$1M$, 89% chance of $\$0$.

Numerous experiments show that subjects typically prefer (a) to (b) and (c) to (d). Let's call such subjects *Allais responders*. Allais responders' preferences violate SEU, because, according to SEU, their behavior would imply,

$$u(\$1M) > u(\$5M) \cdot 0.10 + u(\$1M) \cdot 0.89 + u(\$0) \cdot 0.01 \quad (6.17)$$

$$u(\$5M) \cdot 0.10 + u(\$0) \cdot 0.90 > u(\$1M) \cdot 0.11 + u(\$0) \cdot 0.89, \quad (6.18)$$

which by adding Equation 6.18 to Equation 6.17 yields,

$$u(\$1M) + u(\$5M) \cdot 0.10 + u(\$0) \cdot 0.90 > u(\$5M) \cdot 0.10 + u(\$1M) + u(\$0) \cdot 0.90,$$

which is impossible.

Machina (1989) converted this version of the Allais paradox to a version about making decisions involving two stage gambles. Suppose that in an experiment the subjects consist of Allais responders who would also choose a sure \$1M to a gamble of receiving \$5M with probability $\frac{10}{11}$ and \$0 with probability $\frac{1}{11}$. This is an eminently reasonable hypothesis because Allais responders choose (a) over (b).

Consider Situation 1 where the decision maker has to make a choice between gamble A and gamble B in prospect (F):

(F) There is an 89% chance that the decision maker will receive \$1M and a 11% chance that she will be able to choose between gamble A and gamble B.
- Gamble A: A sure \$1M.
- Gamble B: A $\frac{10}{11}$ chance of receiving \$5M and a $\frac{1}{11}$ chance of receiving \$0.

Note that choosing A is logically the same as choosing a in the static version of the Allais paradox, and choosing B is logically the same as choosing b in that paradox. Because of this, we expect *before playing the gamble* Allais responders would decide they would want A over B if that opportunity were to occur. If *during the playing of* (F) the occasion arises where an Allais responder must decide between (A) and (B), then she would still choose (A) because of the earlier discussion of this point. Thus for prospect (F), an Allais responder is dynamically consistent, because preferences prior to playing the gamble agrees with her preferences while playing the gamble.

Now consider Situation 2 where an Allais responder has to make a choice about the prospect (G):

(G) There is an 89% chance that the decision maker will receive \$0 and a 11% chance that she will be able to choose between gamble (C) and gamble (D).

- Gamble (C): A $\frac{10}{11}$ chance of receiving \$5M and a $\frac{1}{11}$ chance of receiving \$0.
- Gamble (D): A sure \$1M.

Then choosing (C) is logically the same as choosing (c) in the static version of the Allais paradox, and choosing (D) is logically the same as choosing (d) in that paradox. Because of this, we expect *before playing the gamble* Allais responders would decide they would want (C) over (D) if that opportunity were to occur. If *during the playing of* (G) the occasion arises where an Allais responder must decide between between (C) and (D), then she would still choose (D) because of the earlier discussion of this point. Thus for (G), an Allais responder is dynamically inconsistent, because her preferences prior to playing the gamble disagrees with her preferences while playing the gamble.

Allais responders behavior in prospect (G) is similar in many ways to the behavior of subjects in disjunction effect: Before engaging in the prospect, an Allais responder makes a decision as to which one of two alternatives is better. This decision is in conflict with the one she makes when receiving the experience that puts her in the position to actually make the decision between the alternatives. Busemeyer's and Tversky & Saffir's explanations accounting for similar conflicting decisions in the disjunction effect do not work here, because they would depend on an Allais responder having less (or at least different) relevant information available before playing the gamble than during playing gamble. But in this case all the relevant information is available in both situations. Interestingly, the decision maker can even know before engaging in the prospect that she will make a different decision if she gets the opportunity to choose between (C) and (D). In such a situation, an Allais responder may want, before she engages in the gamble, to bind herself somehow to choosing (C) if the opportunity for choosing between (C) and (D) arises, much like Odysseus' choice to bind himself to the mast while passing the isle of the sirens.

It is worthwhile to note that (F) and (G) are identical except for the outcome associated with the .89 probability: In (F) it is \$1M and in (G) \$0. Thus by the above discussion, whether an Allais responder is dynamically consistent when the .11 probability event occurs depends critically on the outcome of .89 event—an event that can no longer occur. Thus the dynamical inconsistency observed in (G) appears to depend on irrelevant information, which from a traditional economic perspective, renders an Allais responder decision process irrational. However, from a psychological perspective, there is more to decision making about gambles than the processing of the dimensions of the gamble's information. There may be other dimensions associated with the dynamics of carrying out the gamble. Emotion is one obvious candidate. It can change subjective experience and thus lead to a revision of utilities or subjective probabilities without changing objective information about the world outside the decision maker. For such situations, both psychology and economics need their modeling to include a decision maker whose internal states change with respect to time. Orthomodular modeling is one way to account for such changes, where the internal states of a subject at different times correspond in a between-subjects experiment to subjects participating in different experiments.

Consider the paradigm consisting of Allais responders as subjects and prospects F and G as experiments. Because the subjects are Allais responders, they choose A in F and C in G. Thus the propositional lattice consists of two propositions,

$$\pi(A) = \pi(C) = 1 \text{ and } \pi(B) = \pi(D) = 0.$$

When Experiment F is invoked, 1 appears as outcome A, and when Experiment G is invoked, 1 appears as outcome C. Thus, *in paradigm space, A and C* describe the same proposition, while *in their respective experiment spaces,* they correspond to different events. Similarly for B and D. But given the prospects F and G, A and D (and similarly, B and C) describe the same event in the physical world. Psychologically, in F, A has some characteristic that makes it a preferable choice to B, and similarly in G, C has some characteristic that makes it a preferable choice to D. These characteristics are psychological in nature and presumably have something to do with the perceived riskiness of gambles. The important consideration is that these characteristics cannot be completely determined in terms of properties of the gambles: they also need information about how they are viewed and interpreted by the subject. Subjectively, this information is attached to the gamble, and not the viewing condition. For example, the gamble in the physical world described by B may be perceived as "much more risky" than the gamble described in the physical world by A, but the gamble in the physical world described by D may be perceived as only "slightly more risky" than the gamble described in the physical world by C, thus making A a more desirable choice in F than B, and C a more desirable choice in G than D.

In summary, F and G can be interpreted as two subjective viewings of the same objective situation (or proposition). Each subjective viewing is interpreted as an event with two outcomes. These outcomes have both objective and subjective interpretations and the subjective interpretations have characteristics not found in objective interpretations. The changing of the viewing from F to the viewing from G changes the outcomes in F into those of G. This does not change objective interpretations, but may change some or all of the subjective interpretations.

6.5 A Use of Probability in Making Logical Inferences Across Experiments

The paradigm's propositional lattice, $\mathfrak{P} = \langle \Pi, \subseteq, \mathbb{U}, \mathbb{m}, -, O, \varnothing \rangle$, can be looked at as providing a logical structure on the set of propositions Π. In some circumstances, probabilistic relationships involving the theory's probability function, $\mathbb{P}$, add additional logical structure that is useful in confirming hypotheses not derivable from the paradigm's theory. This section presents an illustration.

Example Consider the situation of a between-subject paradigm consisting of the following two experiments:

- Experiment 1: The economy has significantly fallen from time t. Each subject in this experiment must decide to either buy property α (Buy$_1$) or don't buy α (Not Buy$_1$).
- Experiment 2: The economy has significantly risen from time t. Each subject in this experiment must decide to either buy property α (Buy$_2$) or don't buy α (Not Buy$_2$).

Because of the nature of the property, the experimenter hypothesizes that Buy$_1 \to$ Buy$_2$, i.e., everyone who chose Buy$_1$ in Experiment 1 would had counterfactually chose Buy$_2$ in Experiment 2. The following will be shown: If $\mathbb{P}(\text{Buy}_1 \mathbb{U} \text{ Not Buy}_2) = \mathbb{P}(\text{Buy}_1) + \mathbb{P}(\text{Not Buy}_2)$, then Buy$_1 \to$ Buy$_2$.

$$(6.19)$$

Note that the truth of the "If" part in Equation 6.19 is decided by the paradigm's data. If it holds, then the experimenter's hypothesis is confirmed.

To show Equation 6.19, assume

$$\mathbb{P}(\text{Buy}_1 \mathbb{U} \text{ Not Buy}_2) = \mathbb{P}(\text{Buy}_1) + \mathbb{P}(\text{Not Buy}_2). \qquad (6.20)$$

Then by the definition of $\mathbb{U}$,

$$\ll \text{Buy}_1 \cup \text{Not Buy}_2 \gg \ = \ \ll \text{Buy}_1 \mathbb{U} \text{ Not Buy}_2 \gg \ .$$

Thus from Equation 6.20 and the definition of $\mathbb{P}$,

$$\mathbb{P}(\text{Buy}_1 \cup \text{Not Buy}_2) = \mathbb{P}(\text{Buy}_1) + \mathbb{P}(\text{Not Buy}_2). \qquad (6.21)$$

The previous equations will be used to show

$$\ll \text{Buy}_1 \cap \text{Not Buy}_2 \gg \, = \, \varnothing . \qquad (6.22)$$

Equation 6.22 follows by first noting that each subject in $\ll \text{Buy}_1 \cup \text{Not Buy}_2 \gg$ is counted exactly once in the computation of $\mathbb{P}(\text{Buy}_1 \cup \text{Not Buy}_2)$ and at least once in either computation of $\mathbb{P}(\text{Buy}_1)$ or in the computation of $\mathbb{P}(\text{Not Buy}_2)$. Note that if s were some subject in $\ll \text{Buy}_1 \cap \text{Not Buy}_2 \gg$ then that subject would be counted once as an element of $\ll \text{Buy}_1 \cup \text{Not Buy}_2 \gg$, once as an element of $\ll \text{Buy}_1 \gg$, and once as an element of $\ll \text{Not Buy}_2 \gg$, that is, counted once in the computation of $\mathbb{P}(\text{Buy}_1 \cup \text{Not Buy}_2)$ and twice in the computation $\mathbb{P}(\text{Buy}_1) + \mathbb{P}(\text{Not Buy}_2)$. This would make $\mathbb{P}(\text{Buy}_1) + \mathbb{P}(\text{Not Buy}_2)$ larger than $\mathbb{P}(\text{Buy}_1 \cup \text{Not Buy}_2)$, contradicting Equation 6.21. Because each subject must choose exactly one of the outcomes Buy_2, $NotBuy_2$, those subjects who chose Buy_1 must choose Buy_2 because $Buy_1 \cap NotBuy_2 = \varnothing$. Thus $Buy_1 \to Buy_2$.

This example shows that the background lattice and orthoprobability theory can provide scientists with tools that may provide for a different and better understanding of empirical data.

6.6   Boolean Subalgebras and Blocks

The following theorem provides another means of using empirical probabilistic relationships involving propositions from different paradigm experiments to derive relationships having the logical and probabilistic structures of classical logic and traditional probability theory.

Of particular interest are boolean subalgebras of the paradigm's lattice of propositions, $\mathfrak{P} = \langle \Pi, \subseteq, \uplus, \sqcap, -, O, \varnothing \rangle$. As discussed below, they provide ways for testing hypotheses about relationships among the paradigm's experiments.

Let $\mathfrak{P} = \langle \Pi, \subseteq, \uplus, \sqcap, -, O, \varnothing \rangle$ be the paradigm's lattice of propositions and $\mathbb{P}$ be its lattice probability function. The following theorem shows that each boolean sublattice $\mathfrak{B}$ of $\mathfrak{P}$ is a boolean algebra of events and that the restriction of $\mathbb{P}$ to Π is a traditional probability function on $\mathfrak{B}$.

Theorem 6.2. *The lattice of propositions,* $\mathfrak{P} = \langle \Pi, \subseteq, \uplus, \sqcap, -, O, \varnothing \rangle$*, is an orthomodular lattice. Suppose* $\mathfrak{B} = \langle \mathcal{B}, \subseteq, \uplus, \sqcap, -, O, \varnothing \rangle$ *is a boolean* subalgebra of $\mathfrak{P}$. Then for all A and B in $\mathcal{B}$, if $A \Cap B = \varnothing$ then

$$A \Cap B = A \cap B, \quad A \Cup B = A \cup B, \quad \text{and} \quad \mathbb{P}(A \cup B) = \mathbb{P}(A) + \mathbb{P}(B).$$

Proof. In Section 6.2 it was shown that $\mathfrak{P}$ is an ortholattice and $\mathbb{P}$ was an orthoprobability function on $\mathfrak{P}$. Thus by Theorem 6.1, $\mathfrak{P}$ is orthomodular. Let A and B be arbitrary elements of $\mathcal{B}$ such that $A \Cap B = \varnothing$. Because $\mathfrak{B}$ is a boolean subalgebra, $A \subseteq -B$. Thus

$$\mathbb{P}(A \Cup B) = \mathbb{P}(A) + \mathbb{P}(B). \tag{6.23}$$

Because $\mathfrak{P}$ is finite and orthomodular, it follows from Theorem 2.28 that $\mathfrak{P}$ is atomic and

$$A = \{\text{the set of atoms in } A\} \quad \text{and} \quad B = \{\text{the set of atoms in } B\}.$$

Because $\cup$ is the least upper bound operator on $\langle \wp(O), \subseteq \rangle$ and $\Cup$ is the least upper bound operator on $\langle \Pi, \subseteq \rangle$, it follows that $A \cup B \subseteq A \Cup B$. It will be shown be contradiction that $A \Cup B = A \cup B$. Suppose $A \cup B \subset A \Cup B$. Because for all distinct outcomes a and b in O, $\{a\}$ and $\{b\}$ are atoms of $\mathfrak{P}$ and $\{a\} \cap \{b\} = \varnothing$, it follows from the definition of $\mathbb{P}$ and an easy generalization of Equation 6.23 that

$$\mathbb{P}(A) = \sum_{a \in A} \mathbb{P}(\{a\}) \quad \text{and} \quad \mathbb{P}(B) = \sum_{b \in B} \mathbb{P}(\{B\}),$$

and similarly, because $A \cup B \subset A \Cup B$,

$$\mathbb{P}(A) + \mathbb{P}(B) = \sum_{a \in A} \mathbb{P}(\{a\}) + \sum_{b \in B} \mathbb{P}(\{B\}) < \sum_{z \in (A \Cup B)} \mathbb{P}(z) = \mathbb{P}(A \Cup B). \tag{6.24}$$

(Equation 6.24 follows because all atoms have a positive probability of occurring and that all atoms in $A \cup B$ are atoms of $A \Cup B$, and $A \Cup B$ has at least one additional atom.) Equation 6.24 contradicts Equation 6.23. Thus $A \Cup B = A \cup B$. Because A and B were selected as arbitrary elements of $\mathcal{B}$, it also follows that

$$-A \Cup -B = -A \cup -B,$$

and thus that

$$-(-A \Cup -B) = -(-A \cup -B),$$

which by DeMorgan's Laws yields,

$$--A \Cap --B = --A \cap --B,$$

which, because $\mathfrak{B}$ is boolean, yields $A \Cap B = A \cap B$. $\quad \square$

A boolean sublattice $\mathfrak{B}$ of $\mathfrak{P}$ is said to be *maximal* if and only if for all sublattices $\mathfrak{C}$ of $\mathfrak{P}$, if $\mathfrak{B} \subseteq \mathfrak{C}$ then $\mathfrak{B} = \mathfrak{C}$. A maximal boolean sublattice of $\mathfrak{P}$ is called a *block (of $\mathfrak{P}$)*. By Theorem 6.2, blocks are boolean algebra of events. It is immediate that $\mathfrak{B}_i = \langle \wp(O_i), \subseteq, \cup, \cap, O_i, \varnothing \rangle$ is a block of $\mathfrak{P}$ for each i in I. There can be additional blocks.

Paradigm experiments can be viewed as contexts for eliciting choices and producing probabilities for outcome events. Viewed this way, blocks provide the same kind of structural information as paradigm experiments—including having a traditional probability function defined on it that is based on empirical data. Thus an intriguing possibility for analyzing and evaluating the theory and data of a psychological paradigm is to not only describe how its experiments are interrelated with respect to its theory and data, but also describe how the paradigm's blocks are so interrelated. Focusing on blocks adds additional structure, perhaps allowing stronger conclusions to be drawn. I currently know of no methodology for this for psychological paradigms as developed in this chapter, but I am confident that one can be developed.

6.7 Other Non-Boolean Modeling of Decision Making

Although context is a much studied phenomenon in psychology, there has been little formal work on the nature of contextual invariance and relationships among contexts. Recently, two new approaches for decision theory have been developed for this. The first, which is based on pseudo complemented distributive lattices, has been described in Chapter 5. The second, and a more influential one, is quantum probability theory (e.g., Busemeyer & Bruza 2012), which is the probability theory used in quantum mechanics stripped of its physical interpretation. The latter is an orthoprobability theory with an event space being a lattice of closed subspaces of a Hilbert space and the orthoprobability function having special features that are important for quantum mechanics. Both of the new approaches have been used to explain violations of SEU and the impacts of context, ambiguity, and vagueness on decision making.

The main problem I see with Hilbert space probability modeling for decision theory and more generally psychology is its lack of an adequate foundation based on psychological theory. As described in this chapter, orthoprobability theory, which generalizes Hilbert space probability theory, meshes nicely with experimental practices and reasoning involving experimental paradigms. However, it lacks the mathematical power of Hilbert space probability theory, and therefore is inadequate for carrying out some of the modeling and theory that can be achieved with Hilbert space probability theory.

Hilbert space probability theory has an underlying rich geometric foundation. I see this as a serious foundational problem for developing it as a rigorous, methodological foundation for decision making. My skepticism results from a lack of an observable—or even a plausible—similarly rich geometric structure in decision experimental paradigms. Orthomodular probability—perhaps with modest enhancements—is far more reasonable for foundational purposes. The challenge for it is to demonstrate—with perhaps modest enhancements—that it can provide an acceptable theory for the kinds of decision phenomena that are currently tackled with quantum probability theory.

References

Allais, M. (1953). Le comportement de l'homme rationnel devant le risque: Critique des postulats et axiomes de l'école américaine. *Econometrica 21*, 503–546.

Birkhoff, G. (1934). Applications of lattice algebra. *Proc. Camb. Phil. Soc. 30*, 115–122.

Birkhoff, G. (1948). *Lattice Theory, 2nd revised edition.* Providence, R. I: Amer. Math. Soc.

Birkhoff, G. and von Neumann, J. (1936). The logic of quantum mechanics. *Annals of Mathematics, 37(4)*, 823–843.

Boole, G. (1854). *An Investigation of the Laws of Thought: on which are founded the mathematical theories of logic and probabilities.* London: Walton and Maberly.

Busemeyer, J. R. and Bruza, P. D. (2012). *Quantum Models of Cognition and Decision.* Cambridge: Cambridge University Press.

Cantor, G. (1895). Beiträge zur Begründung der transfiniten Mengenlehre. *Math. Ann., 46*, 481–512.

Chichilnisky, G. (2009). The topology of fear, *Journal of Mathematical Economics, 45*, 807–816.

Dedekind, R. (1900). Uber die drei Moduln erzeugte Dualgruppe. *Math. Ann., 53*, 371–403.

de Finetti, B. (1972). *Probability, Induction and Statistics.* New York: Wiley.

Dilworth, R. P. (1945). Lattices with unique complements. *Trans. Amer. Math. Soc., 47*, 123–154.

Dirac, P. A. M. (1930). *The Principles of Quantum Mechanics.* Oxford: Oxford University Press.

Ellsberg, D. (1961). Risk, ambiguity and the Savage axioms. *Quaterly Journal of Economics, 75*, 643–649.

Fine, T. L. (1973). *Theories of probability; an examination of foundations.* New York, Academic Press.

Frink, O. (1941). Representation of boolean algebras. *Bull. Amer. Math. Soc., 47*, 755–756.

Foulis, D. J. and Randall, C. H. (1972). Operational Statistics. I. Basic Concepts. *Journal of Mathematical Physics, 13*, 1667–1675.

Fox, C. R. and Birke, R. (2002). Forecasting trial outcomes: Lawyers assign higher probabilities to scenarios that are described in greater detail. *Law and Human Behavior, 26*, 159–173.

Fox, C. R., Rogers, B., and Tversky, A. (1996). Options traders exhibit subadditive decision weights. *Journal of Risk and Uncertainty, 13*, 5–19.

Gilboa, I., Postlewaite, A., and Schmeidler, D. (2009). Is it always rational to satisfy Savage's Axioms? *Economics and Philosophy, 25*, 285–296.

Greechie, R. J. (1971). Orthomodular lattices admitting no states. *Journal of Combinatorial Theory, 10*, 119–132.

Gödel, K. (1931). Über formal unentscheidbare Sätze per Principia Mathematica und verwandter Systeme I. *Monatshefte für Mathematik und Physik, 38*, 173–98.

Gödel, K. (1933). Eine Interpretation des intuitionistischen Aussagenkalküls. *Ergebnisse eines Mathematischen Kolloquiums, 4*, 39–40. English translation in J. Hintikka, Ed., *The Philosophy of Mathematics*, Oxford, 1969.

Heyting, A. (1930). Die Formalen Regeln der intuitionistischen Logik. *Sitzungsberichte der Preussichen Akademie der Wissenschaften*, 42–56. English translation in *From Brouwer to Hilbert : the debate on the foundations of mathematics in the 1920s*, Oxford University Press, 1998.

Huntington, E. V. (1904). Sets of independent postulates for the algebra of logic. *Trans. Amer. Math. Soc., 5,* 288-309.

Husimi, K. (1937). Studies in the foundations of Quantum Mechanics. *Proceedings of the Physico-Mathematical of Japan, 19,* 766–89.

Idson, L. C., Krantz, D. H., Osherson, D., and Bonini, N. (2001). The Relation between Probability and Evidence Judgment: An Extension of Support Theory. *Journal of Risk and Uncertainty, 22,* 227–249.

Luce, R. D. and Krantz, D. H. (1971). Conditional Subjective Utility. *Econometrica, 39,* 253–371.

Kahneman, D., Slovic, P., and Tversky, A. (Eds.) (1982). *Judgment under Uncertainty: Heuristics and Biases.* New York: Cambridge University Press.

Kahneman, D. and Tversky, A. (1979). Prospect theory: an analysis of decision under risk. *Econometrica, 47(2),* 263–291.

Kahneman, D. and Tversky, A. (1982). Judgment of and by representativeness. In Kahneman, D., Slovic, P., and Tversky, A. (Eds.), *Judgment under Uncertainty: Heuristics and Biases.* New York: Cambridge University Press.

Kalmbach G. (1983). *Orthomodular Lattices.* London: Academic Press.

Kolmogorov A. (1933) *Grundbegriffe der Wahrscheinlichkeitsrechnung.* Republished as *Foundations of the Theory of Probability.* New York, Chelsea, 1946, 1950.

Krantz, D. H., Luce R. D., Suppes, P., and Tversky, A. (1971). *Foundations of Measurement, Vol. I.* New York: Academic Press.

Lopes, L. L. (1987). Between hope and fear: The psychology of risk. In *Advances in Experimental Social Psychology.* L. Berkowitz (ed.). New York, Academic Press. 20: 255–295.

Luce, R. D. (1991). Rank- and sign-dependent linear utility models for binary gambles. *Journal of Economic Theory, 53,* 75–100.

Macchi, L., Osherson, D., and Krantz, D. H. (1999). A note on superadditive probability judgments. *Psychological Review, 106,* 210–214.

Machina, M. L. (1989). Dynamic consistency and non-expected utility models of choice under uncertainty. *Journal of Experimental Literature*, *27*, 1622–1668.

Narens, L. (1974). Minimal conditions for additive conjoint measurement and qualitative probability. *Journal of Mathematical Psychology, 11*, 404–430.

Narens, L. (1985). *Abstract Measurement Theory*. Cambridge, Mass.: The MIT Press.

Narens, L. (2003). A theory of belief. *Journal of Mathematical Psychology, 47*, 1–31.

Narens, L. (2005). A Theory of Belief for Scientific Refutations. *Synthese, 145*, 397-423.

Narens, L. (2007). *Theories of Probability: An Examination of Logical and Qualitative Foundations*. London: World Scientific.

Narens L. (2011). Probabilistic lattices: theory with an application to decision theory. In E. N. Dzhafarov & L. Perry (Eds), *Descriptive and Normative Approaches to Human Behavior*. New Jersey: World Scientific, 161–202.

Narens, L. (2014). Multimode utility theory. Manuscript.

Nikodym, O. (1960). Sur le mesure non-archimedienne effective sur une tribu de Boole arbitraire. *C. R. Acad. Sci. Paris, 251*, 2113–2115.

Pedersen, A. P. (2014). Comparative expectations. Manuscript.

Pothos, E. M. & Busemeyer, J. R. (2009). A quantum probability explanation for violations of "Rational" Decision Theory. *Proceedings of the Royal Society B, 276 (1165)*, 2171–2178.

Rasiowa, H. and Sikorski, R. (1963). *The Mathematics of Metamathematics*. Warsaw: Panstwowe Wydawn. Naukowe.

Redelmeier, D., Koehler, D., Liberman, V., and Tversky, A. (1995). Probability judgement in medicine: Discounting unspecified alternatives. *Medical Decision Making, 15*, 227–230.

Robinson, A. (1966). *Non-Standard Analysis*. Amsterdam: North-Holland.

Rottenstreich, Y. and Tversky, A. (1997). Unpacking, repacking, and anchoring: Advances in Support Theory. *Psychological Review, 104*, 203–231.

Saliĭ, V. N. (1984). *Lattices with Unique Complements*. Providence, RI: American Mathematical Society.

Savage, L. J. (1954). *The Foundations of Statistics*. New York: Wiley.

Schechter, E. (2005). *Classical and Nonclassical Logics*. Princeton: Princeton University Press.

Scott, D. (1964) Measurement models and linear inequalities. *Journal of Mathematical Psychology, 1*, 233–247.

Simon, H. A. (1957). *Models of Man, Social and Rational: Mathematical Essays on Rational Human Behavior in a Social Setting*. New York: John Wiley and Sons.

Stone, M. H. (1936). The theory of representations for boolean algebras. *Trans. of the Amer. Math. Soc., 40*, 37–111.

Stone, M. H. (1937). Topological representations of distributive lattices and Brouwerian logics. *Čat. Mat. Fys. 67*, 1–25.

Szász, G. (1963). *Introduction to Lattice Theory (Third Revised and Enlarged Edition)*. New York: Academic Press.

Tarski, A. (1930). Une contribution à la théorie de la mesure. *Fund. Math., 15*, 42–50.

Tversky, A. and Bar-Hillel M. (1983). Risk: The long and the short. *Journal of Experimental Psychology: Learning, Memory, and Cognition, 9*, 713–717.

Tversky, A. and Kahneman, D. (1974). Judgment under uncertainty: Heuristics and biases. *Science, 185*, 1124–1131.

Tversky, A. and Kahneman, D. (1981). The framing of decisions and the psychology of choice. *Science, 211 No. 4481*, 453–458.

Tversky, A. and Koehler, D. (1994). Support Theory: A nonextensional representation of subjective probability. *Psychological Review, 101*, 547–567.

Tversky, A. and Shafir, E. (1992). The disjunction effect in choice under uncertainty. *Psychol. Sci. 3*, 305–309.

von Mises, R. (1928). *Wahrscheinlichkeit, Statistik un Wahrheit*. (2nd ed., 1936, J. Springer. 2nd edition translated into English as *Probability,*

Statistics and Truth, Dover, 1981).

von Neumann, J. (1932). *Mathematische Grundlagen der Quanten-mechanik.* Berlin: Springer.

Weatherson, B. (2003). From classical to intuitionistic probability. *Notre Dame Journal of Formal Logic, 44 (2),* 111–123.

Wright, R. (1990). Generalized Urn Models. *Foundations of Physics, 20,* 881–903.

Zadeh, L. A. (1965). Fuzzy sets. *Information and Control, 8,* 338–353.

Index

Printed in the United States
By Bookmasters